D0819581

THE COMPLETE BOOK OF
RADIO-CONTROLLED
MODELS

THE COMPLETE BOOK OF
RADIO-CONTROLLED
MODELS

How to build and operate model boats, aircraft, cars and trucks

Chris Ellis

CHARTWEL
BOOKS, INC

CONTENTS

A Quarto Book

Copyright © 1999
Quarto Publishing Inc.

All rights reserved. No part of this book
may be reproduced, stored in a retrieval
system or transmitted in any form or by
any means, electronic, mechanical,
photocopying, recording or otherwise,
without the prior permission of the
copyright holder.

Published by Chartwell Books
A Division of Book Sales, Inc.
114 Northfield Avenue
Edison, New Jersey 08837
USA

ISBN   0-7858-0943-0

This book was produced by
Quarto Publishing plc.
The Old Brewery
6 Blundell Street
London
N7 9BH

Senior Project Editor: Gerrie Purcell
Senior Art Editor: Elizabeth Healey
Designers: Julie Francis, Sheila Volpe
Illustrators: Kuo Kang Chen,
    Carol Hill
Photographers: Colin Bowling,
    Paul Forrester
Art Director: Moira Clinch
Publisher: Marion Hasson

Manufactured by Pica Colour Separation
Overseas Pte Ltd, Singapore
Printed by Star Standard Industries
(Pte) Ltd, Singapore

INTRODUCTION

You may have been attracted to radio control by a superb model car at an exhibition, or a realistic model of a fighter bomber being flown in a park, and wondered how to get started. This book can show you how.

GETTING STARTED

First, decide which type of model meets your needs and interests, by reading the introductions to the different types of models in this book (aircraft, pp.38–49; road vehicles, pp.90–101; and boats, pp.146–151). As an introduction to radio control, you may like to start with one of the simple beginner's models recommended on pages 28–35.

You will need to have a suitable space to run your model. Find out where you can operate

ABOVE A good kit for beginners is the Kyosho Inferno DXII, a 4WD off-road racer. It performs well on rough terrain, and many "hot-up" tuning options are available.

models in your neighborhood and check if any local restrictions apply. Parks and recreational spaces, where models are flown, or cars are operated, are good places to meet experienced enthusiasts, who will be ready to offer advice.

Familiarize yourself with the workings of radio control, and its equipment, by reading pages 10–21. Visit hobby stores in your area, and check out the kits, control equipment, and accessories available. Ask if there are any catalogs and brochures that you can take home to study. Local model exhibitions are also worth visiting, since they can be a good source of information on clubs in your area.

USING THE BOOK

Once you have decided on the type of model you wish to build, you can consult the practical projects in this book, (airplanes, pp.50–55, gliders, pp.62–71; helicopters pp.78–83; cars, pp.102–119; trucks, pp.126–135; powerboats, pp.152–163; and yachts, pp.176–179). In each section, detailed instructions and step-by-step photographs show you how to assemble a variety of kit models. Graded with the symbol of a small transmitter, according to the level of difficulty, the models featured here are widely

LEFT The steering and forward and reverse gears of this magnificent Tamiya Ford Aeromax can all be operated by radio control. A 1:14 scale model, it hauls a matching semi-trailer, and is available in metal or plastic.

available from hobby stores. Many similar kits are available, so the variety of models here should give you a good idea of the type of model suited to your skills and experience. If you have any difficulty finding a suitable model, refer to the manufacturers and distributors listed on page 185, who will be able to advise you of your nearest outlet.

If you are new to radio-controlled models, avoid buying the kit you find the most exciting, because it will probably be the most complex. It is better to start with a simple project, such as the Robbe Laser powerboat (p.152), or the Tamiya Fighter Buggy RX (p.102), because finishing and learning to operate a simple model will be far more satisfying than struggling with a more complex kit, and giving up halfway. However, once you have some experience, you will be able to try some of the more advanced projects in this book, such as the Stratus Sports powered glider featured on page 66, or the Kyosho Hyperfly helicopter shown on page 78.

MASTERING THE SKILLS

Before you start to build your model, make sure that you have a good working space and the right tools for the job. A comprehensive list of the tools needed for general model making can be found on pages 12–13. It is also a good idea to familiarize yourself with the basic construction techniques on pages 18–27, before you start.

Your modeling skills will gradually improve if you work through the graded projects in this book. Each model is accompanied by step-by-step instructions and photographs. Following these steps will help you to understand more about the electronic aspects of radio-controlled models, in addition to giving you a chance to practice more conventional modeling skills, such as assembling wooden components.

You will also find many tips on operating your model, once it is completed. On pages 76–77, for example, hints are given on flying a helicopter, a model notoriously difficult to operate. You can improve your car-handling skills by building the circuit suggested on page 98. Helpful diagrams on pages 56–57 show you how to master the art of launching a glider, what to do once it is in the air, and how to get it back without crashlanding.

When you have mastered the basic skills and successfully completed some of the more advanced projects in this book, you may feel inspired to build a model from scratch from your own design. These can vary from simple projects to the complex airplane and boat models seen in major adventure movies. You can find inspiration for future models by taking a look at the Gallery sections, which feature a magnificent array of finished models, ranging from a fully functioning replica of a World War II submarine to a sophisticated award-winning helicopter.

ABOVE Scale replicas of real airplanes are some of the most challenging models to build and operate. This is a scale model of the famous Douglas A-4D Skyraider in US Navy finish.

BELOW A good example of the exciting range of radio-controlled power boats available today is the Graupner Mini Sprint speedboat. This model is electrically powered and fast, yet compact.

BASICS

No matter what type of model you decide to build, you will need basic modeling skills and a knowledge of tools and materials and how to use them safely. This illustrated section describes all the skills you need to know to complete a kit successfully.

Radio control can be confusing at first, but once you have a grasp of the basic elements, you will be ready to build and operate a simple model.

Futaba Skysport 6A is an example of one of the more sophisticated transmitters.

BASICS OF RADIO CONTROL

Although radio-controlled models vary considerably in appearance and size, they all operate on the same basic principles. This is true whether they are car, aircraft, yacht, or powerboat models. Once you have grasped the basic principles and can visualize how everything relates to the control system, then it is much easier to understand the work involved in building and operating a model.

RADIO FREQUENCIES

The radios used to control the models transmit on certain designated frequency bands in the same way as broadcasting stations use designated frequency bands. These, in turn, are different from the frequency bands allocated to public services such as the police or air traffic control. This ensures that none of the allocated frequency bands interferes with the others. Several models can operate at the same time, because the frequency bands are broken into channels, or separate frequencies. Note that the frequency bands allocated for radio-controlled models vary from country to country. In practice the transmitters sold for

This Trainer with its engine running will rev up and roll for take off when the pilot moves the throttle control lever on the transmitter.

The steering-wheel-type transmitter, such as this from the Futaba Megatech range, is favored by some car and boat modelers.

radio-controlled models in each country will already be tuned to the allocated frequencies. However, you still need to check that the equipment you purchased is suitable for your model.

CRYSTALS

A key component of every radio-control system is the crystal (sometimes abbreviated to Xtal). The crystal vibrates at high speed to form and receive the signal transmission. A crystal marked TX is fitted in every transmitter and a crystal marked RX is fitted in every receiver. Each crystal is marked with the channel number, and the two crystals, TX and RX respectively, must each have the same number for the control system to work. If the crystals do not correspond, or if one or both are damaged, there will be no connection between them and control will be lost. Crystals are removable, however, so they can easily be replaced. This flexibility means that you can ensure that no two models operating near each other are on the same frequency band simply by checking and comparing the channel numbers marked on the crystals. You may also wish to alter the frequency band on which the model operates, and you can do this simply by changing the crystal.

Matchbox-sized receivers have made radio-controlled models more practical.

TRANSMITTERS

The transmitter is the hand-held radio set, or control box, which transmits the radio signal to the model. The servos, which control movement in the model, move in proportion to the movement of the control stick on the transmitter. This is known as proportional control. A simple example is a boat's rudder. Move the control stick half to the left and the rudder moves half to the left. Move it full to the left and the rudder will go hard over to the left also. Transmitters may be 2-channel, 4-channel, or more, depending on the model and the degree of control that is necessary. Another type of transmitter is the pistol controller, which usually has a trigger-type control stick. This kind of transmitter is often used for model cars. Some sophisticated transmitters with digital printouts are now available.

RECEIVERS

The receiver is the equipment in the model that receives the radio signal from the transmitter and interprets the instructions for each servo. The receiver is usually controlled by a switch on the model and is powered by a nickel-cadmium, or ni-cad, battery pack.

A typical servo. Note that the shape and size of the output arm may vary.

SERVOS

A servo is a key piece of equipment (usually a small black box with a linkage) in your model. This small part controls movement, typically moving a control surface or adjusting a throttle by means of a connecting rod. There is usually one servo per function on a model. They come in dozens of shapes and sizes, depending on the model. For example, lightweight servos are needed for small or scale models, and special servos for sail winches on yachts. Servos wear out, and consequently you will need to know how to replace them. Associated with the servo is the output arm on the spindle, which may be disk-, star-, or T-shaped. Attached to the output arm is the pushrod or linkage, which transmits the movement to the item being controlled, such as the airplane control surface, an engine throttle, or a car steering arm. Some kits include output arms, but they are also sold separately. Pushrods are sold separately, too, and may need replacing if damaged.

ELECTRONIC SPEED CONTROLLERS

The electronic speed controller, another "small black box," is used to control engine speed in electrically powered models. (In internal combustion motors, a servo-controlled throttle is used to control engine speed.) For the electronic speed controller and the electric motor, another battery pack is required.

NI-CAD BATTERIES

Ni-cad rechargeable battery packs are made in sizes to suit every type of radio-controlled model. Some models with limited space, such as the Hyperfly helicopter, have specially designed packs. A battery charger is essential, since ni-cad batteries have a short life. In practice you will need several ready-charged battery packs on hand during an operating session.

Patriot RV is typical of the wide array of electronic speed controllers available.

SPECIAL EQUIPMENT

Some models have extra equipment to optimize their performance. For example, helicopters have a gyrostabilizer to prevent them from autorotating. A recent innovation is an autopilot for airplanes, which automatically returns all control surfaces to "neutral" so the aircraft can land gently if there is a loss of control.

LEFT Ni-cad battery packs come in numerous shapes and sizes.

ABOVE Battery chargers are essential for recharging ni-cad batteries after operation.

TYPICAL CONTROL EQUIPMENT

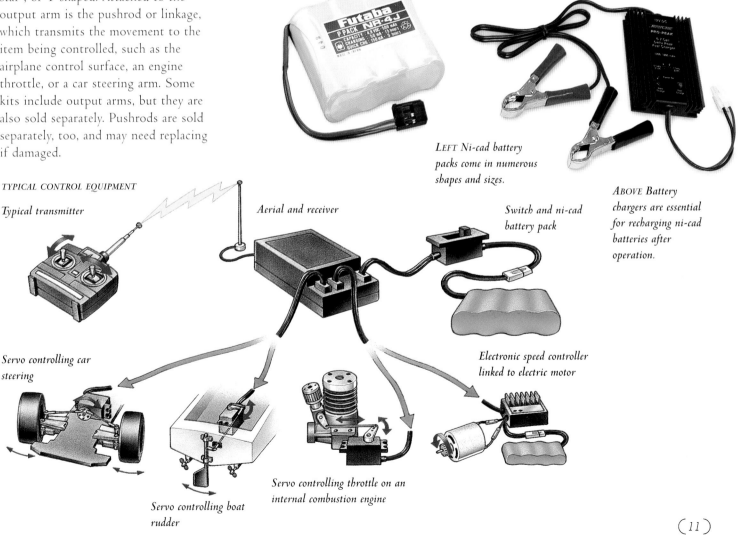

Typical transmitter

Aerial and receiver

Switch and ni-cad battery pack

Servo controlling car steering

Electronic speed controller linked to electric motor

Servo controlling throttle on an internal combustion engine

Servo controlling boat rudder

Basic equipment will be needed for most models. However, you will need some specialist items, particularly if you are making a model aircraft.

TOOLS AND MATERIALS

You will need a set of basic tools and a variety of other equipment and materials to make a radio-controlled model. Some specialist tools will also certainly be necessary for the construction and repair of many models, particularly aircraft and cars. However, it is worth remembering that certain tools, such as Allen keys and wrenches, are supplied in many car kits.

You will need also some specialized materials in addition to the basics. Building model aircraft, in particular, requires the use of special materials. These have usually been developed for a specific task and are particularly efficient.

ABOVE A basic tool kit should contain light-, medium-, and heavy-duty craft knives; an E-ring tool; and an Allen key.

BELOW This Excel tool kit contains a large selection of knives, blades, and small tools.

ABOVE This deluxe Excel tool kit contains everything you need to build a wooden or plastic model.

BASIC TOOLS

If you do home maintenance or other constructive hobbies, you may already have many of the basic tools. Your tool box should include the following: modeling knife and spare blades, razor saw, small screwdrivers, emery boards, small "rat-tail" file, pin vise drill, steel rule (as a cutting edge), tweezers, and flat-nosed pliers.

LEFT Small pliers are useful modeling tools.

SPECIALIST TOOLS

Once you start to make more complex models, you will need to use more specialized tools. Some will only be used for certain types of models, while others will be useful for any kind of model.

A set of Allen keys, or wrenches, in a range of sizes is useful for releasing or tightening inaccessible screws. You will also find liquid thread lock—a glue for screws—useful for making screw-assembled models secure. Nut drivers are nonessential, but a great timesaver, particularly if you are servicing a racing car during a pit stop.

In some advanced aircraft kits, you may have to cover the wing or other surface during assembly. For this work you will need a sealing iron and a heat gun. A small soldering iron is sometimes useful, but rarely required with modern kits. Specialist tools that are specifically designed for use with plastic film coverings, such as cutters, perforators, and trimmers, are also available.

RIGHT This handy Custom sealing iron is used to apply plastic heatshrink film and has full temperature control.

RIGHT The tail parts of some models are made only of wood sheet, as in the case of this P-51 Mustang tail assembly. The complete model is shown on page 44.

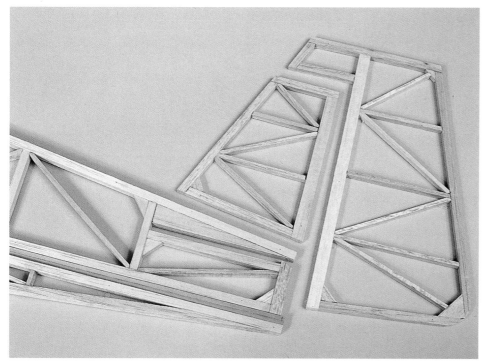

LEFT The tail parts of a Profile Slim model under construction. Because of the flat fuselage, this model is easier to assemble than most other wooden kit airplanes.

SAFETY FIRST

Some tools and materials can cause injury if used incorrectly. Never leave tools or materials unattended, especially when there are children about. If you are using a spray or glue, work outdoors or make sure your work area is well ventilated. For more information on safety, see pages 16–17.

ABOVE Some glues and materials for model work, including PVA glue, shrinking dope, sanding sealer, 5-minute epoxy, thinners, and cleaner.

BASIC MATERIALS

You will need a variety of basic materials in the course of making all but the simplest of models. A selection of glues will be useful for all but the most basic of models. For general work, "universal" adhesives are handy. To secure foam and other plastics, you will need PVA adhesives, also known as white craft glue. Superglue is also a useful general glue, but always keep a debonder handy in case of accidents. If you are making a balsa airplane, you will need balsa cement. Use glue sparingly, for safety reasons and to keep your work neat. For more safe handling tips, see page 17.

Abrasive materials, such as sandpapers and garnet papers, are useful items for your materials storage box. You will find them handy for cleaning down plastic molding marks and also for removing spilled glue. From time to time you may also need a variety of plastic and metal in sheet form and other shapes produced specifically for model making. These materials will come in useful for repairing damage and reinforcing models with structural damage. For repairing or building airplanes from scratch, balsa sheet in all thicknesses, balsa block, and balsa strip are available. If the model is particularly large, you may need to use stronger thin plywood.

ABOVE Here, a coat of dope is being applied to cover a balsa and plywood model with tissue. The tissue panel is cut oversize so that it will wrap around the leading and trailing edges of the wings.

The specialized lubricants used in model making: thread lock, switch lubricant, molybdenum grease, and ceramic grease; and applicator nozzles.

BELOW This scale Fairchild Argus with a wingspan of 50in (1.27m) is made entirely from scratch using balsa wood and a scale plan. The next step for the modeler, Victor Lowe, is to cover the aircraft with tissue.

LEFT Here, an Ultimate Slim model is being covered with plastic heatshrink film.

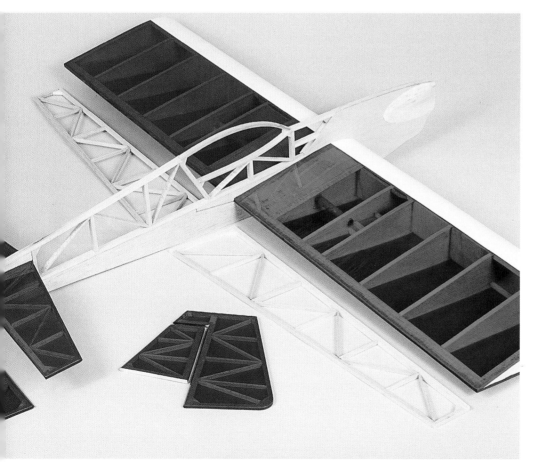

SPECIALIST MATERIALS

As you progress to more complex models, such as IC-engine cars and aircraft, you will need more specialized materials, such as dope and fillers. When you buy a kit, check whether you need to buy any extra materials. Common requirements in more complex aircraft kits include: tissue or film covering in several different grades, weights, colors, and finishes (for wings); epoxy glass and resin (for wing fillets and joints); and dope (for wing coverings). If you are making a model with an IC engine, you will need fuel-proofer; spray oil; and cleaner sprays.

There are several important safety considerations that you should be aware of before starting any model project.

SAFETY

Many of the techniques you will use to assemble your kit will be the same whatever type of model you choose. The same tools, glues, and paints used in making static models will also be useful in the construction of radio-controlled models. However, if handled carelessly they can be dangerous. Always make sure you understand how to use all tools and materials and read any instructions carefully before you get immersed in building your model. To avoid causing injury, ruining your model, or making costly mistakes, follow the safety guidelines below both when building and when operating your model.

KEY SAFETY RULES
Don't be put off radio-controlled models by fears about safety—if you follow these rules you are unlikely to have many problems. If any friends or family members are helping you, always discuss safety with them before you begin.

Make a safe work area
You may be lucky enough to have a toolroom, a workroom, or a den where you can build your models. Most enthusiasts, however, are limited to the kitchen table, or a table in the yard or on the porch in warm weather. Before you start, make sure that your work won't disturb anybody else in the household. You will also need to prepare a suitable work surface. Choose a rigid, portable surface, such as thick plywood, chipboard, or composition (MDF) board, at least 2ft x 2ft (60cm x 60cm). A newspaper under the board will prevent scratching and also catch any accidental oil, paint, or glue spillage. A cutting mat that you can put on your work surface is useful if you need to do any serious cutting or sawing.

Make sure that your work area is well lit. Good lighting, such as a portable lamp with an adjustable, flexible stand, is needed not only for accurate modeling but also for safety reasons.

Avoid working in shadow or with your back to the main room light. If you work indoors in daylight, face the window or other light source and try to avoid any shadows in your work area.

Keep your tools safe
As you gain experience in model building, you will build up a sizable collection of small tools, many of them capable of causing serious injury if not handled correctly. It is imperative to get a tool box big enough and strong enough to hold all your tools safely. Hardware stores sell purpose-built tool boxes, with compartments and a secure lid and carrying handles. Fishermen's boxes, with their many compartments, are an adequate substitute. Always return all tools to the tool box when not in use, and when you finish a work session, check your work area carefully to make sure you haven't left any tools out.

Use your tools safely
Many injuries are caused by careless or accidental handling of tools. Use the different compartments of your tool box for separating your tools and keeping them tidy. In particular, keep knives and saws away from your other tools. Old film containers or small screwtop jars can be used to store small drillbits, tacks, nails, or knife blades until you need them.

Craft knife blades and small saws are potentially the most dangerous modeling tools of all. The best way to avoid injury to yourself or others is simply to use them with care and concentrate on your task. Always cut away from you, and if you are holding a small kit part with one hand, ensure that your fingers are nowhere in the cutting line. If your knife hits a tough part while cutting, don't try to force it. See if you can cut from another angle or use a bigger knife. There are several different types of modeling knife, including craft knives with fixed exposed blades, and safety knives with disposable blades.

ABOVE Use a cutting mat well illuminated from above, so that all cutting can be done safely. Here, a craft knife is being used against a steel rule to cut a plastic card. Note that the fingers are kept clear of the cutting line.

ABOVE RIGHT It is essential to wear safety glasses while drilling. Here, a metal support strut is being drilled to insert a bolt. The work area is well illuminated, and the fingers are kept well clear of the drill.

ABOVE FAR RIGHT Check all electrical or control components before using. Set out the ni-cad battery and charger and identify the leads before connecting them up.

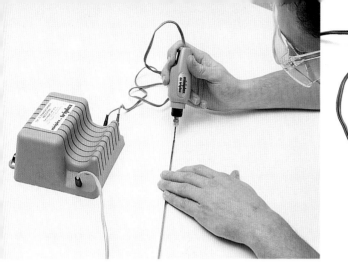

Sharpest of all are frisket knives (scalpels), which are available in model stores. When you replace the blade, never handle it directly—it is best to use a small pair of pliers. Always dispose of broken blades or knives carefully by wrapping them in newspaper to prevent injury.

If you are doing any drilling, grinding, heavy filing, heavy sanding, power sawing, or other work that generates dust or fragments, always wear safety glasses to prevent injury or irritation to the eyes.

Look after the kit parts

Losing a part can hold up progress or even prevent you from completing the model. Lost parts can also be a hazard to small children. Open the small component bags or bubble packs only when you need the parts, and place them on shallow box lids (shoebox lids are ideal) to stop them straying.

Take care with the electrics

Be extremely careful when handling the electric parts of your model, because you can damage the parts (and also yourself with burns and electric shocks) if you short a circuit or join up the wires incorrectly. Electric tools, such as soldering irons or small drills, should be handled carefully. Always make sure you are holding a drill properly before you switch it on, to avoid injury.

Avoid paint and glue spillage

You will need to use a variety of glues and adhesives which are potentially hazardous if not handled correctly. The CNA superglues are particularly dangerous because they bond in seconds. Other glues give off noxious fumes and can cause headaches or other problems if your work area is not well ventilated. Always remember to put the lids back tightly on paint and oil jars to avoid inadvertent spillage or leakage of fumes. Paints, oils, and glues can also be a fire hazard if

not stored correctly, so keep them in a secure box.
Always keep a stock of debonders, thinners, and spirits to hand so that you can cope with any spillage at once. PVA glues (also known as white craft glue) are water soluble, so any PVA spillage can be cleaned up with a wet cloth. "Universal" glues are spirit-based and can be cleaned up with thinners or mineral spirits.

Keep a first-aid kit

It's a good idea to keep a basic first-aid kit at hand in case you cut yourself while modeling. Several sizes of Bandaid dressings, some lint or cotton, tweezers, disinfectant, and antiseptic cream are all useful for light nicks and cuts. If the injury is more serious—a deep cut or a severed fingertip—get medical help immediately.

Supervise children

Never allow children unlimited access to your work area or the models themselves. Most kits are intended for serious modelers and are not recommended for anyone under 14 years. However, ten-year-olds (or younger) can handle the simpler RTR (ready-to-run models)—a selection appear in this book—with adult supervision. With the exception of IC-engine cars, these models require no complicated assembly work and need only batteries to operate.

Operate in safety

Be aware that a radio-controlled model not under proper control, or operated carelessly, can cause injuries to others, resulting in litigation in extreme cases. It is a good idea to take out insurance to cover not only damage to the model but also inadvertent injury to others. For more information and advice, consult an insurance broker or check if your local club has an insurance policy.

SAFETY FIRST

For particular safety requirements for each type of model—cars, boats, aircraft, and trucks—see the appropriate section.

● Make a safe working area.
● Use your tools with care.
● Always return your tools to the tool box after use.
● Take care when assembling the electrics.
● Ensure your hands and surroundings are completely dry before handling electrical equipment or making connections.
● Keep the kit components in a safe place.
● Keep a first-aid kit to hand.
● Supervise children.
● Always keep debonders and cleaners handy.
● Take great care when testing your model during assembly.
● Never operate, even for test purposes, without first giving a warning.
● Always make sure there is enough room to maneuver the model.
● Take out insurance to cover damage and also inadvertent injury to others.

You will need basic modeling skills to assemble a radio-controlled kit. However, many of the building procedures are common to all models.

BASIC CONSTRUCTION TECHNIQUES

Although it is possible to make radio-controlled models entirely from scratch using raw materials, and servos and control equipment purchased separately, beginners and many average modelers will choose to build a model from a kit. Radio-controlled model kits can be quite complex, and you will need basic modeling skills to complete them successfully. Once you have chosen a kit appropriate to your experience, take time to study the instructions. Whatever the type and level of model you choose to build, the most important tip for successful model construction is to take your time—always think ahead and plan your next move carefully.

ASSEMBLY

The quality of kit components can vary widely. Some models are made to such high engineering and molding standards that they screw or clip together easily. Other kits may require more preparatory work, such as drilling screw holes

using a template, or strengthening molded plastic parts with wood strips. This may be due to manufacturering limitations with vac-form moldings.

Most kits with wood parts are now much easier to build than in the past. Wood parts made from balsa or thin plywood are usually die-cut, so only minimal trimming is needed to part the pieces from the sheet. Never cut out any wood parts before you need them, to avoid losing or damaging the pieces.

If you experience real difficulties due to the kit design, consult the hobby dealer who sold you the kit. Some kit manufacturers may also have a helpline for advice and information.

BELOW LEFT Thin plywood frames from a modern aircraft kit.
BELOW RIGHT The fuselage is shaped with a covering of thin balsa sheet.
BOTTOM Thin ply is used for the ribs and spars.

DOS AND DON'TS OF MODELING

- Check that you have all you need in your tool kit and paint kit to complete the project.
- Resist the temptation to open the packages of tiny components. Each pack usually contains all the parts you need for a specific stage of the project, so it is important not to mix up the parts.
- Always read the instructions carefully and check that all the components listed in the instructions are in the box. If the diagrams are small, it is a good idea to make enlarged photocopies of each diagram.
- Always follow the assembly sequence shown in the instructions. However, in some modified kits you may find that the instructions have not been altered to match. An improved motor, for example, may require you to drill new holes or add strengthening pieces. Always look for any notification of production changes in the kit that will invalidate part of the instructions.
- Keep your work area neat and tidy (see page 16). Place small parts on shoebox lids to prevent them from rolling away, and use tweezers to handle tiny components. Place newspaper or white lining paper under your work area to make it easier to find dropped parts.
- Identify all the parts and tools you will need before starting the next stage of work.
- Check the instructions ahead of the stage you are working on. For example, the painting details at the end of the instructions may advise you to paint fuel-proofer around the engine area of an IC-engine car. However, by the time you reach this stage the parts that need fuel-proofing may be no longer accessible.

CONTROL EQUIPMENT

Due to developments in electronics, the control element of radio-controlled models has become less complex. Most models can now be controlled with a "black box," and in almost all cases the electrical connections simply plug or clip together, and you are unlikely to need a soldering iron to make any connections. You will usually have to buy the control equipment separately, since model kits do not include the transmitter, receiver, and servos. Kits are made for the international market, and control equipment cannot be included because of the differing regulations concerning, for example, transmission frequencies, in each country. However, model manufacturers make the choice easier for beginners by recommending appropriate radio-control equipment for the model. The kit and control equipment may sometimes be offered together, sometimes at a specially discounted rate. The Sky Surfer featured on page 32 is a good example of a beginner's model that comes as an all-inclusive package.

However, if you are interested in customizing your model or building a model from scratch there is a huge variety of transmitters, servos, and receivers to choose from in the manufacturers' catalogs. For example you can buy a more advanced transmitter, or improve a model by adding different servos. Different makes of transmitters, etc., are generally interchangeable as long as the specifications and size are the same. The same applies to makes of batteries.

For more on the workings of radio control, see page 10.

ENGINES

There are two main types of engine: internal combustion (IC) and electric. In the world of radio-controlled models, IC engines are often referred to as "engines," and electric engines as "motors," although this distinction is not made in this book.

IC engines

IC engines are used for model aircraft, cars, and powerboats. The same type of engine is used for all types, but with certain modifications. For example, marine engines are usually water-cooled and have a water jacket around the cylinder case. IC engines are often known as glow, or nitro engines. Although diesel and gasoline (petrol) were used in the past, a special fuel containing a mix of methanol and nitro-methane is now more usual. You will find this type of fuel at hobby stores in addition to some specialist fuels such as fuel formulated for model helicopters and high-performance models, and diesel fuel mixes for diesel engines.

Most IC engines are 2-stroke, having a firing stroke every revolution. They are quite simple and have a port for timing rather than a valve. The most common method of porting is known as Schnuerle porting, a term often seen in specifications. Most car and powerboat engines have a pull start that operates a recoil spring to turn the crankshaft and fire the cylinder. With model aircraft, this is done by turning the propeller by hand or with an electric starter motor, although some larger models have pull-start engines. Four-stroke engines for high-performance racing or aerobatic

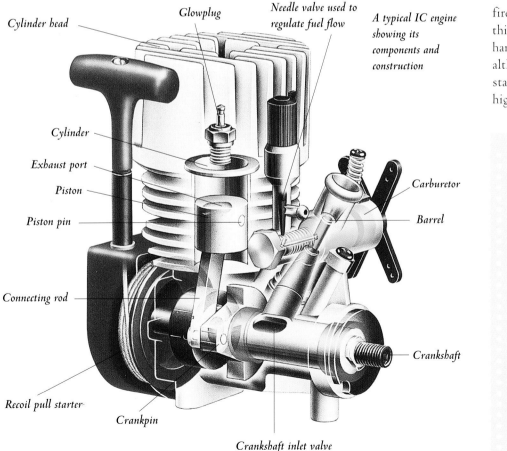

Cylinder head

Glowplug

Needle valve used to regulate fuel flow

A typical IC engine showing its components and construction

Cylinder

Exhaust port

Piston

Piston pin

Carburetor

Barrel

Connecting rod

Recoil pull starter

Crankpin

Crankshaft

Crankshaft inlet valve

BEFORE YOU BUY

• Spend time looking at all the kits on offer in your local hobby store, and don't be afraid to ask for advice.
• Study all the kit manufacturers' catalogs and brochures you can find.
• Always choose a kit appropriate to your experience.
• Make a straightforward model before you try a more complicated kit—operating a basic model successfully is more satisfying than failing to complete a more elaborate kit.
• Always allow for extra items that may not be included in your kit, such as batteries, IC engine (particularly in airplanes), and ballast weight for yachts.

An MDS 38 aero engine, showing the integral muffler.

sport models are also available, but these are more expensive, because they have more moving parts and valves.

IC engines come in various sizes indicated by the cylinder capacity expressed in cubic inches or cubic centimeters. The range of engine sizes available includes: .10, .15, .20, .25, .30, .40, and .60 cubic inches. You can also find bigger engines—such as .75, and variations such as .49, and .61. Kit manufacturers usually specify the size of engine required—in some cases you will have a choice of two sizes, depending on the performance you are seeking. Note that less power is usually required for running a road vehicle. For example, a typical Trainer aircraft or a small helicopter takes a .40 size engine and a large 1:7 scale model might need an engine up to .90 cubic inches, but a car, on the other hand, usually has a .10 or .15 engine.

Although modern IC engines are very efficient and well made, they still require a certain amount of running in and

testing before installation. Test stands are available for this, or the engine can be clamped to a workbench. Once you have started a new engine (you may need several attempts), keep it running for at least 15 minutes, varying the throttle position between idle and full power. For details of how to test an aircraft engine, see page 47.

Most engines are supplied with mufflers, but a variety of special mufflers, such as "ultra-quiet" mufflers and those designed specifically for high-performance models, are also available separately. These have helped to reduce the noise problem on many IC-engine models.

Glow plugs and electric starters also come in many forms and sizes. Special silicone fuel lines and a wide variety of fuel tanks are available. Their fitting relative to the engine position is important, so always refer to the kit instructions. Due to technological developments, improved equipment is appearing all the time. One recent innovation is the Kyosho QRC (quick reverse clutch) gearbox. This is attached to the drive shaft to give full forward and reverse drive

Kyosho QRC clutch (interior) for use with IC engines in cars.

Electric motor installation on a Slim Hellcat model.

operated by one of the control channels. Similar types of gearbox have been developed for electric motors.

ELECTRIC MOTORS

Although they lack the rugged attraction of IC engines, electric motors are increasingly popular, since they overcome the environmental objections associated with internal combustion. Electric motors are quiet, require much less preparation and maintenance, are easier to operate, easier to replace, easier to install, and are easier to gear. Because electric motors are easier and cheaper to make, many radio-controlled models are sold together with an appropriate motor.

Some simpler models, powerboats in particular, have direct drive from the electric motor, but most models have some sort of gears. On more sophisticated models, a proper gearbox, giving forward and reverse gears, is common. A variety of motor sizes is available, including micro size for very tiny light models. The best sizing system for motors was developed by Graupner, a leading manufacturer of

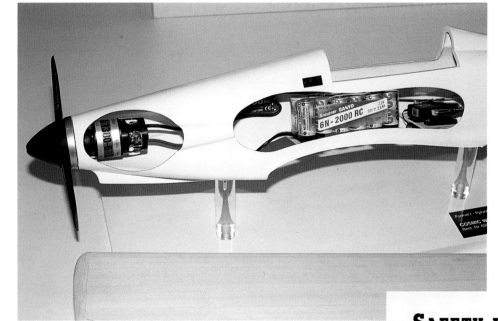

This cutaway shows a top-of-the-range Ultra 1300 electric motor and battery pack in an advanced Graupner Cosmic Wind pylon racer.

motors. In their Speed motor series, motor size is related to case length—for example, a Speed 550 motor has a 55mm case, a Speed 400 motor has a 40mm case. An aircraft would take a 400 motor with gearbox, while large powerboats or trucks would need a 550 motor.

Almost all electric motors are engineered in the same style: they have a cylindrical outer casing housing magnets with a wound armature revolving inside on plain or ball bearings. A condenser, brushes, and commutator are incorporated. Specialized motors include jet power units for speed boats, motors with integral folding propellers for thermal gliders, and outboard motors for powerboats. Increasingly popular are electric impeller, or fan-duct, motors for jet aircraft. With these motors, the propeller is housed in a nacelle, or within the jet pipe of a jet airplane, giving the model an extremely realistic appearance. The Kyosho T-33 Thunderbirds aircraft, for example, uses a Speed 400 motor, giving an 882lb (400kg) thrust. Although this is not a gas turbine motor as in a real jet aircraft, it gives a realistic jet effect.

Simple electric impeller on a Nikko ready-to-fly Jet Delta model.

SAFETY RULES FOR ENGINES AND MOTORS

• Make sure bystanders are kept away from running propellers.

• Take care with fuels for IC engines—they are flammable and toxic. Clean up any spillages at once. Only mix fuel or fill fuel tanks outdoors, using fuel filler bottles with specially designed delivery tubes. Make sure all fuel containers and tanks are securely closed.

• Never use naked lights or electrical equipment in a fueling area. Keep a fire extinguisher to hand.

• Clean and check all engines after each operating session, and check all connections—fuel pipes, electric leads, gearboxes, etc.

• Wear protective stalls on the fingertips to prevent injury when flicking aircraft propellers to start IC engines.

• Avoid touching terminals or bare leads on electric motors, to prevent shocks. Replace any worn leads and cover any bare points with insulating tape.

• Do not touch an engine immediately after running—it will be hot.

• Always recharge batteries after an operating session, and take reserve ni-cad batteries to the operating site.

• Good maintenance and keeping the engine clean will prolong the life of your engine and make your model safer to operate.

Expert painting and finishing skills can give your model a more realistic look. This section shows you all the techniques you need to know to complete the projects in this book.

PAINTING AND FINISHING

BASIC EQUIPMENT

- Paints in basic colors appropriate for the material you are painting (see below).
- Brushes in sizes 2, 6, and 10 (see also page 23).
- Thinners and brush cleaner suitable for the type of paint you intend to use.
- An old cloth (or a supply of paper towels) for cleaning up.

Most of the paints and finishing materials available at hobby stores can be used on any model. However, some paints are specifically formulated for use on radio-controlled models. As with other aspects of model making, it helps to follow proven procedures, as you can easily get into a terrible mess if you do not look after paint in the correct way and use it properly.

TYPES OF PAINT

Several different kinds of paints can be used in model making. Both enamel-based paints and acrylic paints are suitable for styrofoam, wood, metal, glass, and most common hard plastics. They cannot be mixed together, although they do cover each other. This means that you can paint a car body in enamels and then, when it is fully dry, paint over the detail parts, such as the fenders or wheel rims, with acrylics.

Enamels: These are smooth in texture and must be mixed thoroughly before use. Sold in small cans, they are available in a wide variety of colors, including metallic and "ghostly" shades useful for customized cars and aerobatic airplanes. Also useful is clear varnish, available in both gloss and matte. Always use spirit-based thinners and cleaners to avoid damaging your brushes.

Paints can constitute a fire risk and a fume risk, particularly if they spill, so it is a good idea to keep them in a box with a lid and plenty of room for the paint containers to stand upright.

Acrylics: The more modern equivalent of enamels, these are available in a wide variety of colors in both matte and gloss. Sold in plastic containers or screw-top bottles, acrylics should be stirred thoroughly before use. Matte and gloss clear varnish, and appropriate thinners, are also available. Always clean your brush with water after each application, because acrylics dry quickly. They can be easily wiped off while still wet.

Primary colors   *Primary colors plus white*   *Primary colors plus black*

Secondary colors

A basic paint set for model work should contain black, white, red, blue, and yellow paint, in both matte and gloss. You can also mix other shades using these basic colors. A useful addition is an earth or dark earth shade for toning down bright colors or suggesting dirt or rust. You can make colors look faded by adding a few drops of black or gray to light colors or white to dark colors. Before starting to paint, always try out colors on scrap plastic, card, or wood to see how they look.

You will need brushes in different sizes for the range of painting tasks necessary to complete your model.

Polycarbonates: These are specially formulated paints derived from acrylics. They give a good solid coverage on the clear plastic polycarbonate (Lexan) body shells used in many modern radio-controlled car and some helicopter kits. Colors tend to be restricted to the brighter shades associated with cars. Brushes can be cleaned in water.

Dopes: These cellulose-based paints with their distinctive odor are frequently used on the tissue or other material covering wooden aircraft and boats to give a tight fit and provide a hard protective coating. The finish is rubbed down and given a primer coating, after which any paint can be used. Always use cleaners and thinners formulated for use with dopes.

Specialized paints: Solarlac, a modern equivalent of dope, is used for covering the lightweight plastic film used in place of tissue to cover model aircraft with a traditional rib and former construction. Available in a wide range of colors, Solarlac is quick-drying and fuel-proof, so that the painted finish is protected from fuel spillage or discoloration from the exhaust. It comes in gloss only, but you can use a flatting agent to achieve the matte finish needed for camouflage. Fuel-proofer paint in matte or gloss is also available for protecting the area around the engine and exhaust.

APPLYING THE PAINT

There are several different ways of applying paint. You can simply paint your model with a brush, but for a sleeker finish, try using spray paint or an airbrush.

Painting with a brush

You will need a number of different sized brushes for the painting jobs on a single model. Brush size 0 is handy for painting delicate areas, such as a dummy pilot's face; size 2 is good for fine work on most models; larger sizes, such as 5, 7, and 10, are useful for general work. Brushes come in two styles: pointed and flat (chisel edge). The latter is useful for "cutting in" colors, when an area is painted a different color from an adjacent area, as with a duo-tone car finish.

The best-quality brushes are the camel hair, sable, and squirrel types, and properly looked after these brushes can last for years. Always clean your brushes with the appropriate cleaner after use. Restore their correct shape and let them dry out before returning them to your paint storage box.

To achieve a painted surface free of any obvious brush marks when using enamels or acrylics, brush backward and forward in successive straight strokes without lifting the brush from the surface. If you need to paint a second coat—for example, when the first coat fails to cover a light-colored plastic base—allow the first coat to dry thoroughly. An alternative for covering a light-colored base or a transparent Lexan body shell is to use a primer or undercoat in gray before the top coat.

You may want to paint wear, weathering, and dirt effects on your models, and you can do this very realistically with a brush in a variety of ways. For example, you could paint rust effects along the seam and rivet lines on the hull of a model fishing trawler by running some reddish-brown paint into the area while the top coat is still wet. (This is a rare occasion where you might break the drying rule noted above.) You could also use the "dry-brushing" technique to give a less clean finish. First leave the paint finish to dry out completely. Choose the colors you need for wear or weathering (for example, aluminum for scratched paintwork or dark earth for mud), dip in a small chisel-edge brush, and then work out the paint on a piece of cardboard or plastic until there is hardly any paint left on the brush. Now paint the model with light strokes, so that just enough paint is left to suggest wear and tear. For example, paint could be brushed randomly on the lower edges of a rally car to suggest dirt thrown onto the body.

Always leave paint to dry thoroughly before applying another coat. Here, a second coat is applied by brush to the rear spoiler of a GT car.

To use a simple airbrush, attach one end of the air line to the airbrush and the other to the air can. The can of air can be replaced when empty.

The paint reservoir has been filled, and now the airbrush is attached to the reservoir using a special spanner supplied with the kit. All the joints must be airtight.

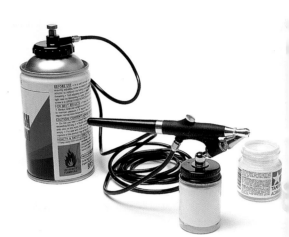

The complete airbrush kit is now ready for use. If possible, use acrylic paint that is specially formulated for airbrush use. The paint reservoir bottle can hold a small bottle of paint.

DOS AND DON'TS OF PAINTING

- Always replace the lid on the paint after use to prevent fumes escaping.
- Always store paint upright to prevent leakages.
- Turn your paint cans occasionally to prevent the pigment from settling.
- Keep paints away from fires and naked lights and keep a fire extinguisher handy.
- Make sure your painting room is well ventilated to prevent buildup of fumes.
- Keep paint away from the eyes. Clean up spilled paint immediately.
- Leave models to dry in a dust-free area, such as the bathroom.
- Make sure your painting area is well lit and never paint in shadow.
- Keep all paints, brushes, and thinners in a covered storage box in a cool place.

Here, the trunk lid of a model car is being spray painted. The area to be covered is masked off and a sheet of paper is used to prevent the paint from marking adjacent areas.

Spray paints

You can get a very professional finish with spray paint, provided it is applied correctly. It is particularly useful for covering large areas, such as a model shell. Once the base paint has been allowed to dry out, smaller areas or details can be painted on with a brush. Spray paint is available in a 4oz (100ml) or 6oz (150ml) aerosol can, which is generally enough for at least two car body shells. Although spray paint does not come in as wide a range of colors as enamels or acrylics, all the popular colors and shades are available. Before you start to spray, cover the surrounding area with old newspapers to catch the "fall out" from the spraying

process. It is best to spray your model outdoors or under open cover on a calm day, but if this is not possible, make sure that the room is well ventilated. Begin by reading the instructions on the can and make sure you know which way it will spray by checking the directional arrow and nozzle position. Hold the can about 12in (30cm) from the model when you spray, but always check the recommendations as some paint manufacturers may advise differently. For best results, swoop smoothly from side to side, keeping steady pressure on the spray nozzle. Hesitant spraying can lead to blobbing or spattering rather than a smooth coverage.

Airbrushing

Airbrushing is considered to be the best way to achieve a perfect, professional finish on a radio-controlled model. It is particularly practical for covering large-scale models. Modern technical innovations have made the process of airbrushing much easier than it was, and a variety of airbrush equipment and paints is now available. The simplest types consist of an airbrush with a paint reservoir attached to a can of spray propellant. At the other end of the scale are airbrushes that work off a small electric compressor, and those that use an internal-combustion compressor. Whatever their type, all airbrushes operate on the same principle: a fine jet

Before using the airbrush, mask off the areas to be covered. Paint can be applied with precision, because the airbrush nozzle is easy to control.

Masking tape is a very useful material. It is best to use the tape intended for models rather than office or domestic brands.

You can use masking tape to paint fine decorative markings on your model. Here, a fine brush with acrylic paint is being used to paint a decorative stripe on a car trunk.

Clear masking fluid can be used to protect surfaces such as windshields and windows when painting or spraying. The masking fluid sets as a gel and can then be rubbed off after painting is finished, together with any accidental paint spills.

of compressed air is forced past a paint reservoir, paint is taken with it and then projected onto the surface to be painted in a very fine spray by way of a small nozzle. The most sophisticated airbrushes have a variable nozzle allowing a strip 1in (25mm) wide to be painted in a single pass at the highest setting, or a fine line only 1/32in (0.8mm) wide at the lowest setting. Airbrush paint must be thin, and some paint ranges are formulated specially for airbrush use. Others, such as the Tamiya acrylic, can be used for both regular brushes and airbrushes. Acrylic paint can be thinned to a suitable consistency by using paint thinner.

The main drawback of using airbrushes is that they have to be taken apart and cleaned thoroughly after use to prevent them from becoming clogged with old dried paint.

MASKING, TRIMMING, AND FINISHING

Use masking tape to paint adjacent areas in different colors. Paint the entire panel in the first color and allow to dry. Then mask off the area to be painted with masking tape, and apply the second color up to and over the edge of the tape. When the second color is perfectly dry, peel off the masking tape carefully to reveal a neat, straight edge. If possible, use the masking tape specially produced for modelers, which is available in assorted widths, because it is thinner and less sticky than regular masking tape. If you need to mask off a curved area, try using flexible masking that can be "bent" to follow a curved line.

You can decorate models with fancy trim lines, or apply "go faster" stripes on competition models, using masking tape, or a roll of thin pre-colored film produced specifically for this purpose.

Available in many popular colors in a variety of widths, the trim can be applied right onto the paintwork or film covering. Some types are self-adhesive, while others have to be secured in place with a coating of clear varnish.

Masking is an important technique in painting radio-controlled car body shells molded in tough, transparent polycarbonate (Lexan). These shells are painted on the inside, giving a very realistic gloss finish. You can also achieve realistic flush windshields and windows by carefully masking off the window areas so that no body paint gets on to them. Ordinary masking tape can be used, but a gel-like masking fluid is often preferred for this task. Available from some paint makers, such as Humbrol, the gel is applied over the window areas and allowed to set a little before you start to paint. When the

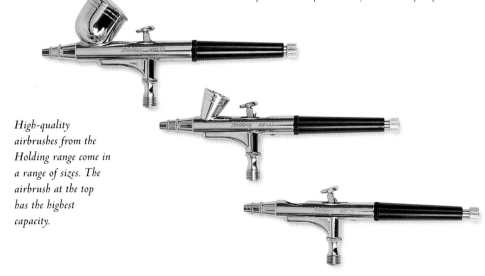

High-quality airbrushes from the Holding range come in a range of sizes. The airbrush at the top has the highest capacity.

This large-scale model of the Top Flite P-47D Thunderbolt in 1:8 scale demonstrates the impressive effect of perfect painting and markings. The model is so well finished it could be mistaken for the real thing.

body has dried, the masking fluid can be rubbed away, leaving perfect, clear windows. Masking fluid can also be used for other intricate areas, such as helicopter fuselage shells and boat tops with enclosed cabins.

Paint markers are extremely useful for finishing tasks. These look like regular marker pens but they come in the same range of colors as standard paints. They are ideal for painting small details, such as car door handles, and for reaching otherwise inaccessible parts. They can also be used for retouching scratch marks on damaged models.

Raised details, such as door handles, hinges, and other molded parts in plastic can be picked out with a fine brush. For example, the door handles on a car can be painted in black or aluminum. Other details, such as grilles, cockpit framing, and decorative coamings, can be painted in the same way.

Some boat, yacht, and simpler ARTF aircraft moldings are already colored, reducing your painting task quite considerably. In addition, self-adhesive decals are used for the trim lines, so very little painting is required.

DECALS

The final task in completing a model's decor is to add the markings. Most models have markings of some kind. For example, a car may carry license plates, sponsor stickers, model names, and driver names, and an airplane may have military markings or civil registration. Kits usually include all the markings you need in the form of decals. These are either waterslide markings or, more often, very thin self-adhesive decals on backing sheets. Occasionally you may find dryprint pressure decals supplied with your kit. If you want to convert the model to something different from what the kit maker intended, or simply to change the letters on a car license plate, or alter the national markings on an aircraft, you will need to buy extra decals. Alternative decals are made by some companies that manufacture radio-controlled models. For example, Wedico makes several sheets of markings for radio-controlled trucks. You can also adapt static-model decals available in various scales from hobby stores.

Waterslide decals

To apply a waterslide decal, cut out the marking from the sheet and place in a saucer of water. The decal should be loose in about half a minute. Position the marking on the model, and gently slide the backing paper from between the decal and the surface of the model. Use a cloth or paper towel to press the marking into position, and soak away the excess water. To fix the marking so that it looks as if it is painted on, use a special decal securing solution, such as Solvaset. Available at the bigger hobby

Paint markers are useful for retouching paintwork, since they come in all standard paint colors. Here, a marker is being used to retouch a defect in the painting of a GT car body.

To apply a self-adhesive decal, first cut the required marking carefully from the sheet.

Position the marking carefully, and then press down lightly, taking care to avoid any wrinkles.

stores, Solvaset is a clear liquid, which is painted over the decal and then left to dry thoroughly. Avoid the temptation to touch the decal while it is still wet, because this will ruin it. If Solvaset is unavailable, apply a thin coat of matte or gloss varnish, whichever is appropriate.

Self-adhesive decals

Self-adhesive decals are commonly used for radio-controlled models, since they stand up to rough handling and are suitable for large areas. Some models, such as racing cars, may have a large number of decals. To apply a self-adhesive decal, simply peel it carefully from the backing sheet and place in position on the model. Rub the decal down with a damp, slightly soapy cloth, working outward from the center to expel all the air bubbles. You should find that the decal fits snugly over the molded surface of the model. It is a good idea to give the decal another rub down an hour or two after the first application as this often makes it fit even more tightly over the molded surface. If a self-adhesive decal shows signs of lifting or peeling off later on, spread a thin layer of white PVA glue under the decal and hold the decal down with a damp cloth until it is firmly in place.

This colorful model needs no painting at all! The hull of the Wavemaster motorboat is molded in the base color of white, and the extensive decor and the cockpit windows are all self-adhesive decals.

Dryprint decals

Dryprint pressure decals are rarely used on radio-controlled models because they are so easily rubbed off. However, they are available in a large variety of useful styles and sizes. Decal sheets of numbers and letters are particularly useful for making new vehicle license plates and for registrations and company names. You can improve adhesion by covering the dryprint decal with clear varnish, using more than one coat if necessary.

Hand-painted markings

If you are a talented artist, you will be able to paint superb markings, lettering, and lines on models. However, even if you are not so gifted, you can paint some basic designs using a combination of techniques, such as masking and dryprint. For example, it is quite easy to paint an "Ace of Diamonds" squadron marking on a large-scale Seahawk jet fighter. First mask off a rectangle with tape and paint it white. Then mask off a diamond shape in the center and paint it red. To finish this effective but simple marking, take a suitably sized "A" for Ace from a dryprint decal sheet and position it in the top left and bottom right corners. Although made entirely by hand, this design is easy because only straight lines are needed.

Fix the decal in place by rubbing down hard with a damp, slightly soapy cloth. Rub from the center to the edges to eliminate any air bubbles.

The decal is now in place, together with all the other markings, which have been applied one by one in exactly the same way.

A typical self-adhesive decal sheet from a radio-controlled airplane kit. The markings are very accurate but there are no spares, so decals must be applied correctly the first time.

If you have no experience of kit building or model assembly, don't be put off radio control by the bewildering variety of complex radio-controlled kits available. There are several ready-to-operate models that are a great introduction to radio control for the complete beginner.

YOUR FIRST MODELS

• *Ready to run—no assembly required*
• *Safe and easy to operate*

Fortunately the complete beginner does not have to start with an assembly kit to experience the fun of radio-controlled models. Several manufacturers offer good-quality ready-to-operate models that are just as well engineered as assembly kits, and often cheaper. Ready-to-operate models are ideal for the complete beginner as they will soon give you a good idea of how radio control works. Although you won't get the excitement of assembling a kit, you will still need to prepare and test your ready-to-operate model before it can be used. Depending on size and sophistication, ready-to-operate models have the same sort of transmitter, receiver, and servos as kit-built models, though they may be slightly simplified. Many beginner's ready-to-operate models are smaller than kit-built models, but this can be an advantage, since they can be used in the backyard or house driveways rather than in public parks or open spaces. They also meet the safety requirements that make them suitable for children.

CHOOSE FROM THESE MODELS
Cars
Buggies
Powerboats
Airplanes

Choosing a ready-to-operate model
There are many models on the market that will get you started. All the models shown here are quality products and can be operated just like a kit-built model. Some models sold in chain stores or markets are like toys and may not be very robust.

1 Jeep Wrangler by Nikko
This model is a good example of a well-engineered, ready-to-operate off-road vehicle. Made to 1:16 scale, 12in (300mm) long, 7.7in (192mm) wide, and 6.8in (170mm) tall, the model has strong independent suspension on all wheels and runs well over rough ground though it only has rear-wheel drive. It has a top speed of 12.5mph (20kph). The kit comes with full operating instructions, a transmitter, and a battery charger. To operate, you will need a Mini 7.2V ni-cad battery pack and a 006P 6F22 battery for the transmitter.

Check that the model is complete and properly assembled and that the wheels and steering move smoothly. Insert the 7.2V battery pack under the chassis and clip the wire leads to the connector inside the chassis. To do this, remove the baseplate of the chassis and then clip it back into place. Fit the other battery in the transmitter. If the circuit is working, you will see a small pilot light on the transmitter box.

2 *The model is now ready to operate. The left-hand stick controls forward, stop, and reverse. Move the stick forward to go forward, and backward to go backward. To stop, move the stick to the upright position. The right-hand stick controls the steering wheel. If the model fails to respond, check that the ni-cad battery is properly inserted and connected or that it is fully charged, using the charger supplied with the model.*

Aerial

Dummy steering wheel

Pilot light

2

3 Atlantic 95 motor cruiser by Nikko

This 1:30 scale model of a modern motor cruiser, just 10.8in (270mm) long and 4.4in (111mm) in beam, is one of the smallest and simplest ready-to-operate radio-controlled models available. It is the perfect model if you are interested in boats, or just want a fun model to find out how simple radio control works. Although intended for ponds or lakes, this model is small enough to operate in your bathtub. Despite its size, it has full function radio control, the left-hand control stick controlling forward, stop, and reverse, and the right-hand control stick controlling the rudder. The model uses six AA R6 batteries under the cockpit floor, and a 006P 6F22 battery in the transmitter; these must be purchased separately. There is a simple on/off power switch alongside the cockpit. It has a top speed of 4mph (6kph) and is fully waterproofed.

3

Working rudder

4 Stingray speedboat by Nikko

If you want fast performance, the Stingray is a large ready-to-operate radio-controlled boat with a top speed of 8mph (13kph). Built to 1:12 scale, the boat is 20.6in (520mm) long and 5.8in (145mm) in beam. The model comes complete with transmitter and battery charger, but you will need to buy a 9.6V ni-cad battery pack for operation and a 006P 6F22 battery for the transmitter. To insert the ni-cad battery, unclip the top of the cockpit and place it through the cockpit, as shown. Then clip the battery leads to the connector in the cockpit.

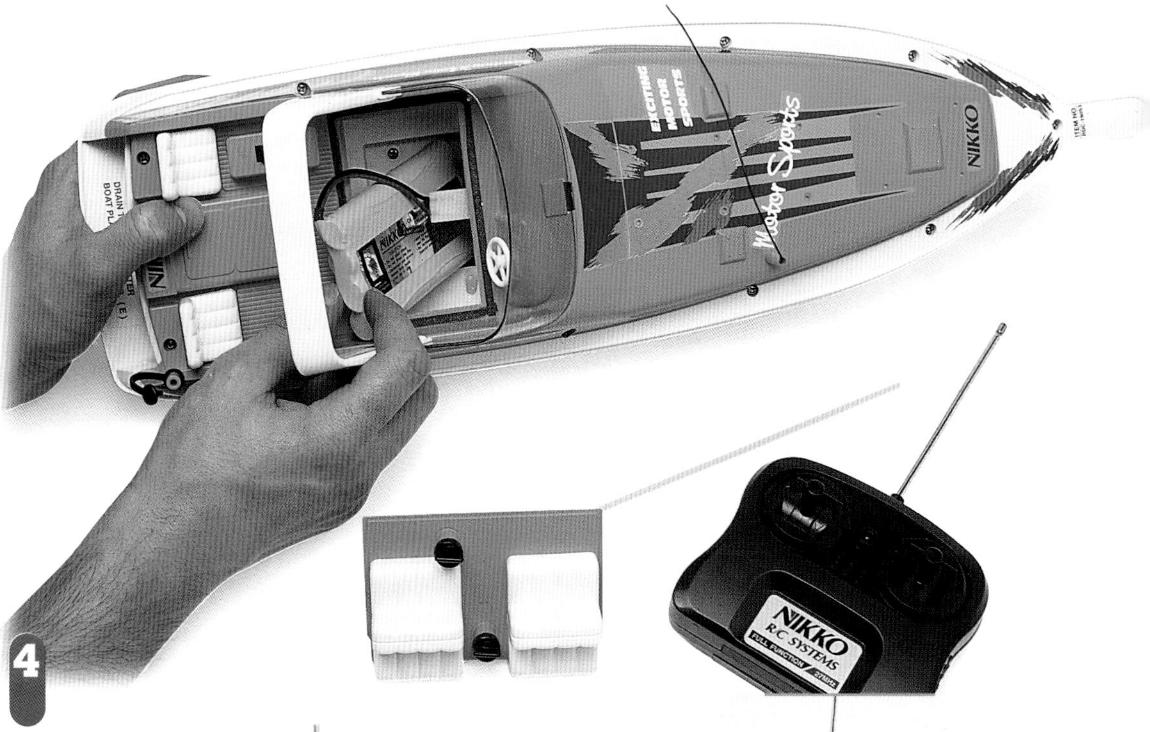

4

5 *It is also necessary to fit the radio aerial so that radio signals can be received. You will find the aerial socket on the front decking. Once the batteries are in place and the aerial fitted, the model is ready to go.*

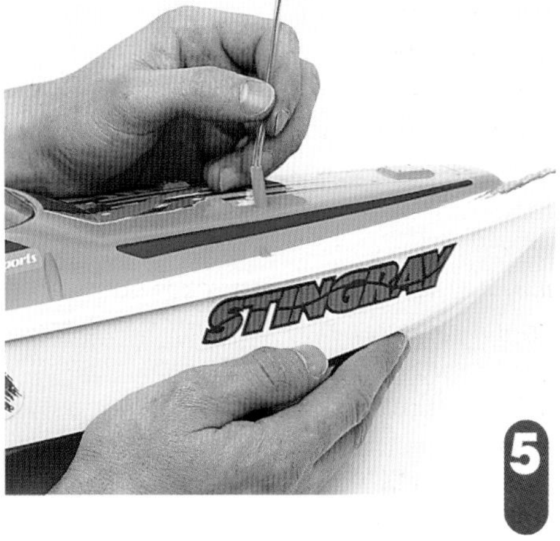

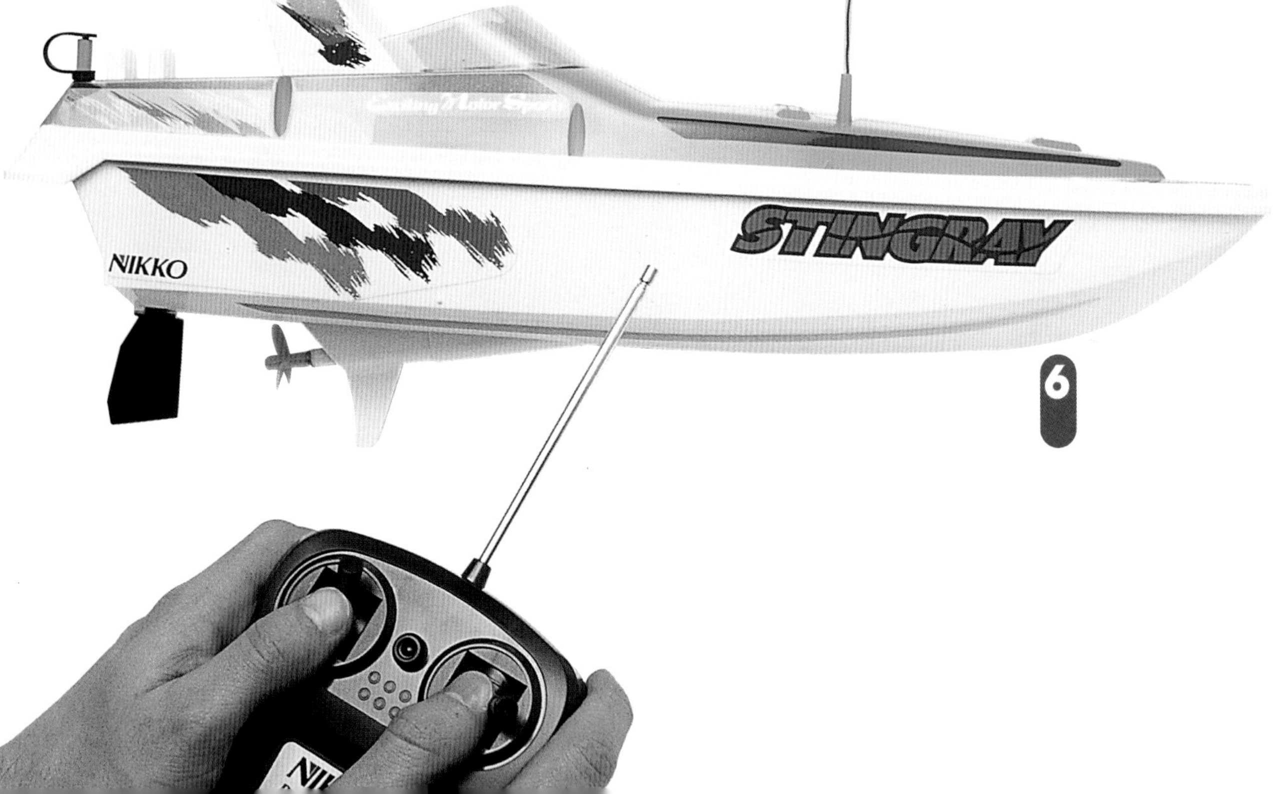

5

6 *As with many models, the left-hand control stick controls forward, stop, and reverse, and the right-hand control stick controls rudder movement. Most models are designed for operation in freshwater; running them in polluted or salt water may cause corrosion and should be avoided.*

6

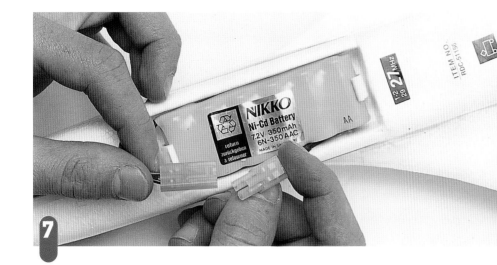

7 Delta Hawk by Nikko

Simple ready-to-fly radio-controlled aircraft, such as the Delta Hawk, are not as widely available as cars and boats. This model is built to 1:14 scale (nominal) and has a wingspan of 24in (600mm) and a length of 29.2in (740mm). It is made of styrofoam moldings for lightness and is very neatly finished. With new batteries it will achieve a top speed of 31mph (50kph). Because it is more advanced than most beginner's models, it comes with a video which shows you how to prepare and fly it. A 7.2V ni-cad battery pack is needed to operate this model, plus four AA R6 batteries for the transmitter. Fit the battery pack into the slot in the bottom of the fuselage and connect the leads to the built-in wiring inside the aircraft.

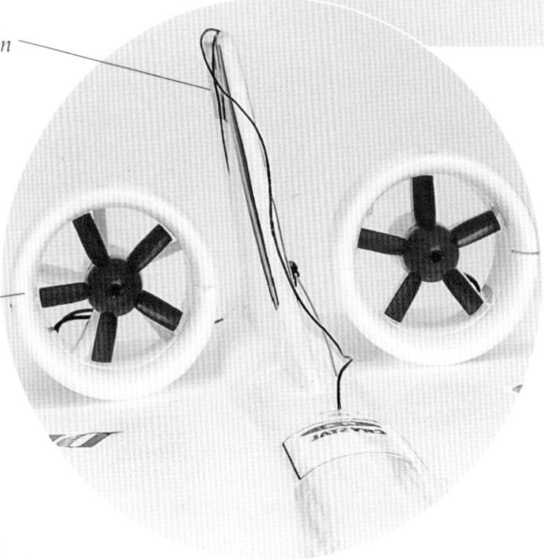

8 The Delta Hawk has a number of special features. These include start switches (for the motors) already fitted in the fuselage, and electrically driven impellers (propellers) concealed inside the engine nacelles and driven from electric motors in the wing roots.

9 The model is launched by hand; the transmitter controls the engine speed and the rudder. Flight duration is 2 minutes—it is a good idea to fly the plane over long grass to minimize landing damage.

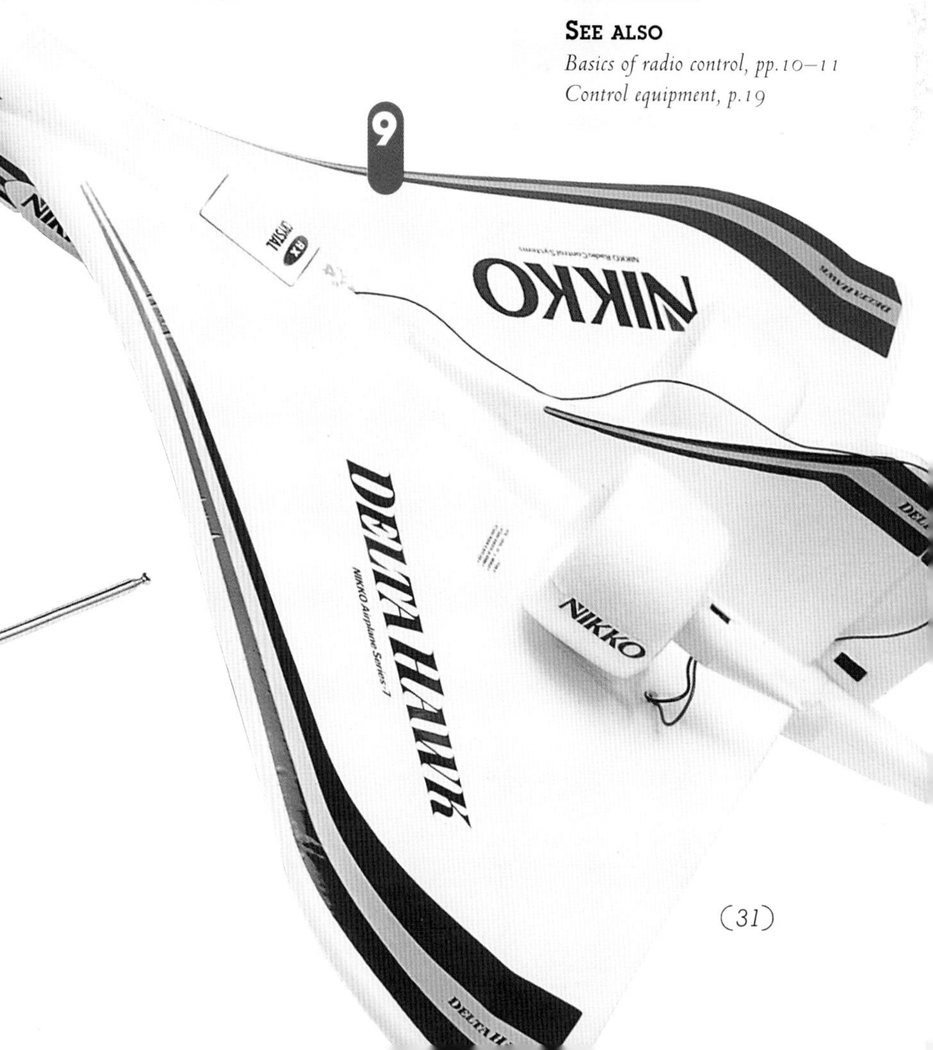

Tail fin

Impeller (ducted fan)

All these good quality models shown are complete, but you may find that models bought in chain stores or markets do not have full function control. Before you buy, check that the model has full function or two-channel radio control.

SEE ALSO

Basics of radio control, pp.10–11
Control equipment, p.19

The Sky Surfer is an ideal project for fledglings: the kit contains everything you need to make and fly the model in one afternoon.

A FLYING START

- ### • Easy to assemble
- ### • Easy to fly
- ### • Control equipment included

Making your first ever radio-controlled model from an assembly kit is bound to be an exciting experience, but one that can be all too easily ruined by choosing a kit that is too complicated for the complete beginner. You can avoid disappointment by choosing a simple, foolproof kit that contains all the components needed to assemble and fly the model. Fortunately there are a number of radio-controlled flying kits available which are aimed at the complete beginner. Try the Sky Surfer by G-Con, as shown here, or similar kits by other manufacturers, such as Robbe's Skyflex.

The Sky Surfer

Unlike many kits which include only the aircraft (or other model), the Sky Surfer kit contains all the components you need to make up the model and fly it, including the control equipment. The only extras you will have to buy are the batteries to power the motor and the control unit.

Another helpful feature of the Sky Surfer is that the complex metal fuselage and the control equipment (which is already installed) come preassembled, so that beginners don't have to worry about these complicated areas.

TOOLS YOU WILL NEED
Small screwdriver

OTHER REQUIREMENTS
Six AA batteries for control box/transmitter
One ni-cad 7.2V battery pack for aircraft
Battery charger

1 Before you start, read the instructions carefully. Check that the kit contains all the main components listed and illustrated in the instructions. Identify the main fuselage assembly (gondola) and take it from the box. Fix the wheels and axles to the fuselage by snap-fitting them in place. The pilot figure shown here is a separate item in the kit.

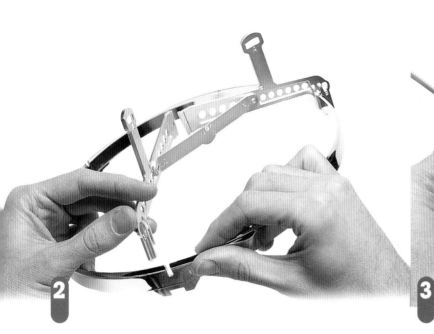

2 *Next assemble the propeller guard and its supports. These are plastic moldings which snap-fit together into molded slots. Ensure that the cross-support is tight and rigid by checking the screws that hold it in place—you may need to tighten them up a little.*

Propeller guard supports

3 *Two servo rods, left and right, are supplied. Slot these over the servo horns which project each side of the gondola. It is essential to support the servo horns with your fingers while the rods are fixed in place to avoid any bending or distortion of the horns. Push up on the servo horns as you press the slotted ends of the rods into the horns.*

4 *Pass the servo rods through the two supporting projections on the propeller guard at the same time as you affix the propeller guard to the main body. Slots on the propeller guard slide into locating shoes on the body. You will hear a click if the correct location is found. If you don't, the part is incorrectly attached and you will need to try again.*

5 *Now take the canopy out of its packing and identify the front suspension lines (green) and the rear suspension lines (pink). Clip the front suspension lines to the eyes in the tops of the left and right servo rods, using the spring-clips that are already fitted. Take the rear suspension lines through slots in the end of the propeller guard supports and attach them to the center holes in the servo rods. If it is a very calm day, these suspension lines can be moved to the upper holes on the servo rod as explained in the flying instructions supplied with the model.*

6 *Find the support rod supplied with the kit and thread it through the holes in the leading edge of the canopy to ensure it fills with air and is correctly shaped in flight.*

7 *Now you can affix the decorative Sky Surfer stickers and the warning signs for the propeller guards. Check that all the rigging is correct and that nothing is tangled or crossed. To prevent this, the suspension lines are bound in tape, but these tapes must be removed before flying the model.*

8 *Finally clip the pilot into the cockpit. Check that the kit is assembled correctly, that the wheels turn freely, and that the propeller is unobstructed. While you are doing this it is a good idea to charge up your 7.2V ni-cad battery so that it is ready to install.*

9 *Now install the batteries. Six AA type batteries fit into the transmitter and a 7.2V ni-cad battery into the housing under the gondola. To fit the ni-cad battery, locate the connector in the battery compartment by opening a hinged door on the underside. Clip the ni-cad battery to the connector, slide it into place under the cockpit, and close the door. You will hear a click if the door is properly closed. You are now ready to test the motor and controls and take the model out for its first flight.*

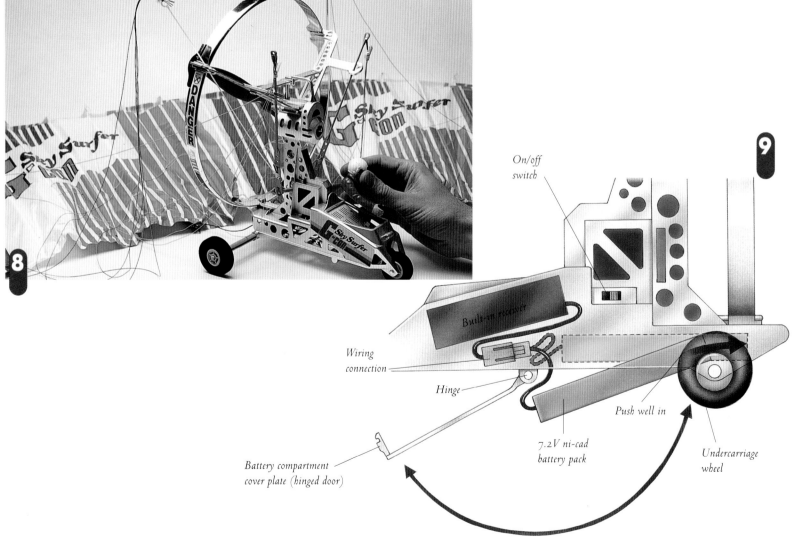

On/off switch

Built-in receiver

Wiring connection

Hinge

Battery compartment cover plate (hinged door)

7.2V ni-cad battery pack

Push well in

Undercarriage wheel

FLYING THE SKY SURFER
It is possible to make and fly the Sky Surfer in a single afternoon. This small model is quiet enough to fly in a park or large yard without many trees. Since it is a simple model it is best to choose a calm day for its first flight. The instructions include a comprehensive section on flying which you should read when you have assembled your model. Take care not to tangle the canopy suspension lines before the launch. It is a good idea to have an assistant for the first few flights.

SEE ALSO

Basics of radio control, pp.10–11
Control equipment, p.19
Introduction to aircraft, pp.38–39

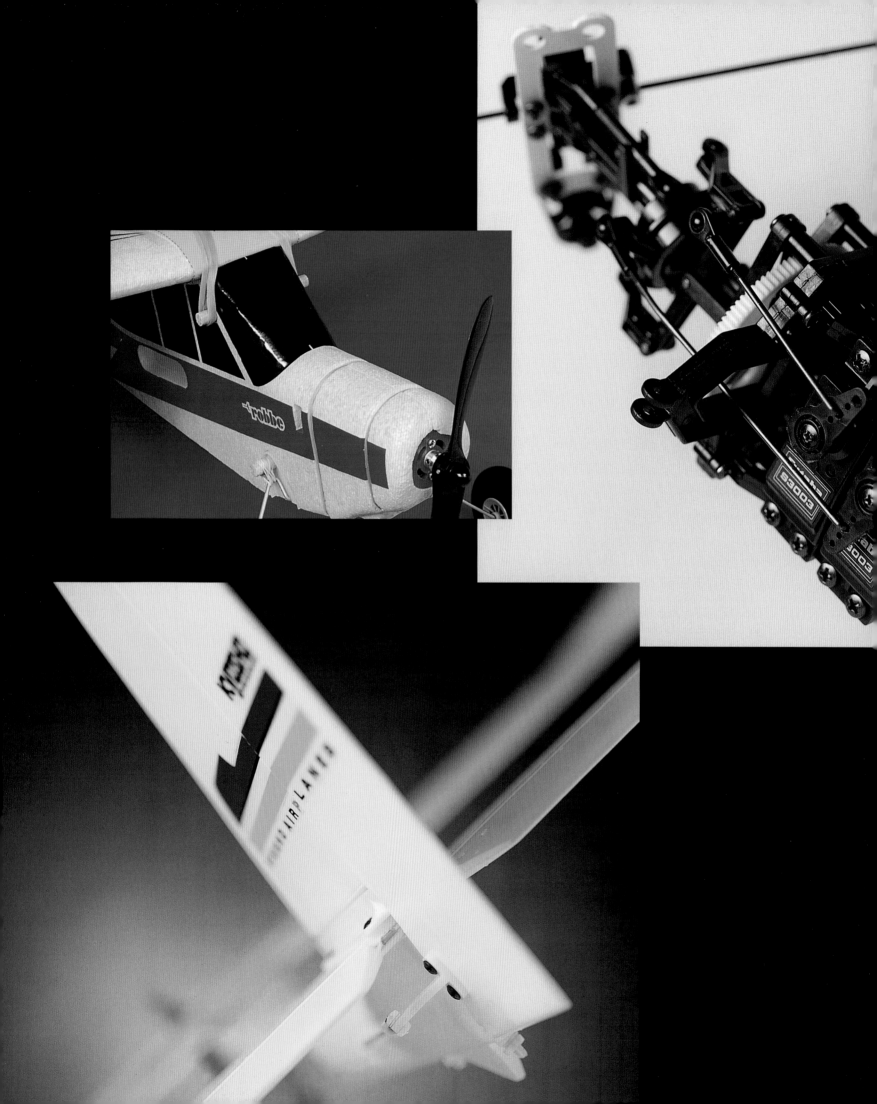

AIRCRAFT

Making a radio-controlled aircraft that will fly and perform well is a challenging yet satisfying task. This section concentrates on basic aircraft kits and the skills that you will need to build and fly them successfully.

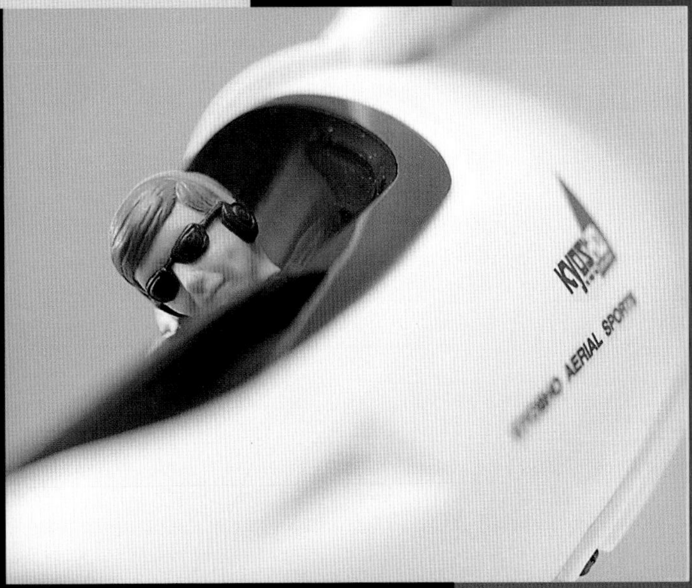

The first radio-controlled models to be developed were airplanes. Today, you can choose from a wide variety of radio-controlled model aircraft. As well as being popular with model enthusiasts, they are frequently used in period dramas and for battlefield reconnaissance.

INTRODUCTION TO RADIO-CONTROLLED AIRCRAFT

Model enthusiasts were inspired to build radio-controlled model aircraft by the development of radio-controlled target aircraft, or drones, which were used by the military for firing practice.

In the 1930s some experimental radio-controlled model aircraft were built, but these were big to accommodate bulky radio equipment. The 1950s saw a rapid development in model aircraft as radio equipment became smaller and the transistor replaced the valve. Today, many different types of aircraft are available in model form, ranging from simple gliders to semi-scale helicopters and true-scale replicas of warplanes, such as the Top Flite P-40E Warhawk, shown inset on the right.

Model aircraft are in demand not only by model enthusiasts but also by the military and the media. The military have found them useful for battlefield reconnaissance: a radio-controlled airplane equipped with a camera can be sent over enemy lines to transmit pictures back to the command center. Film production companies find it cheaper, safer, and more practical to use radio-controlled aircraft, particularly if old aircraft are required. Many of the spectacular bombing raids, air collisions, crashes, and dogfights you see in period movies are actually filmed with model aircraft, which shows just how realistic radio-controlled models can be.

RIGHT The Kyosho Piper Cub J3 is so realistic that it could almost be mistaken for the real thing.

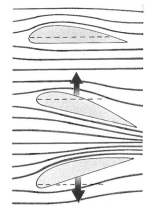

An aircraft's wing creates lift by interacting with the air passing over it. Moving the control surfaces enables the aircraft to climb or dive.
TOP Level flight
MIDDLE Climbing
BOTTOM Diving

GETTING STARTED

Seeing a radio-controlled model airplane in flight may well inspire you to try a model aircraft kit. There is something magical in seeing a model flying just like the real thing, then realizing it is being controlled by someone with a transmitter on the ground. There is a huge number of radio-controlled aircraft enthusiasts today: model-aircraft magazines are by far the most numerous and there are many dealers that specialize in aircraft. As a result many different types of radio-controlled model airplane kits are available, with new kits produced every year.

It is important to choose a kit that is appropriate for your experience. The sight of a sleek fighter plane or a graceful biplane may have captured your imagination and encouraged you to start a kit, but these types of plane are extremely complex. It is far better to choose a simple model, so that you can get an idea of construction techniques and basic flying skills. You will need great skill and patience to build a model aircraft, because they need to be extremely light and fragile in order to fly. You will also have to learn to be philosophical about power or control failures, as it is possible to spend hours on a superbly detailed fighter plane, or an elegant giant sailplane, only to see it wrecked on its first outing.

Once you have mastered the basic construction and flying skills, you can try a more ambitious model. Many challenging aircraft kits are available, ranging from intricate biplanes to the latest high-tech aircraft. If you want to build an aircraft from scratch, model plans and balsa wood

The Fairchild PT-19 by Kyosho is one of the simplest scale models available.

This exploded view of an Airvista Trainer model shows the typical parts found in almost-ready-to-fly (ARTF) kits.

ABOVE The electric-powered Cessna 177 Cardinal M36 is a good example of a modern ARTF model in the Trainer class.

LEFT Scale fighters, such as this superb Top Flite P-51B Mustang, are tempting for beginners, but the simpler Trainers shown on this page are a better choice.

are available from hobby stores. In addition, the leading manufacturers offer a wide variety of accessories ranging from finger guards for handling engines to devices for checking propeller balance on high-tech models.

This section includes examples of simpler radio-controlled aircraft. These will help you to improve your construction and flying skills and prepare you for the more complex model at the end of the section.

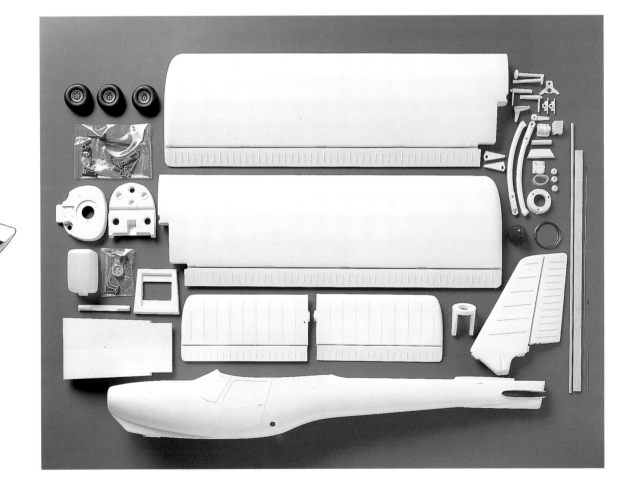

The Ready Trainer kit contains all the parts you need to make a fine model. The aircraft is made of plastic with ABS molded components.

AIRCRAFT KITS

There are several different types of powered aircraft model. Most models are not built to scale, since the proportions need to be changed to improve the model's flying characteristics. Compared with a real aircraft, a model's wings may be wider, the shape simplified, or the wing dihedrals increased. However, some models look more realistic than others—Great Planes make a Trainer that closely resembles a Porterfield M36. True-scale models look just like the real thing, with the exception of their larger propeller, which is necessary for flying.

BELOW The Flash 36 was designed by the model F3A World Aerobatic champion Tsugutaka Yoshioka and features a blow-molded ABS fuselage. Kits are available with either IC or electric engines.

RTF models

There are a few ready-to-fly (RTF) models which are ideal for the youngster or complete beginner. All you need to do before launching is fit the battery pack in place and test the controls. Sometimes you may also have to clip or join the wings and tail to the fuselage, but no actual construction is required.

Beginner's models

These models are ideal for newcomers who want a simple, foolproof model which they can build and fly. Beginner's models generally require some construction work, but if you follow the instructions carefully, you are guaranteed a fully functioning model. Assembly can be completed in a few hours, so there is less chance of losing enthusiasm as frequently happens with complex projects which may take weeks or months to complete. The Airdancer and Sky Surfer featured in this section are just two examples of good beginner's models.

Trainers

High-wing light aircraft, known as Trainers, are the easiest kit-built models to fly. Although some small Trainers are available, the ideal wingspan for stable flying is in the 60–75in (150–190cm) range. Easy to work on, this size of plane shows up well over long flying distances. Some kits are made of wood and require covering in the

The ARTF Kyosho Space Walker is a good example of a semi-scale Sports aircraft. Ninety percent of the model is pre-assembled in the factory.

ABOVE The Autokite II is a simple model, which comes in kit form and features the unusual Rogallo fabric kite wing.

traditional way (see page 23), but most modern kits are "almost-ready-to-fly" (ARTF). With these kits the main structure, including the covering, is pre-assembled. Some designs, such as the Ready, have molded plastic parts which form a monocoque structure of very light weight, similar to a real modern light aircraft. No wood or covering material is required at all. Other kits, such as the AirVista manufactured by Great Planes, simply snap together: all holes are pre-drilled, and everything can be screwed or slotted together in a single evening, and no elaborate tools are needed.

Like real light aircraft, Trainers are generally easy to fly because of their high-wing configuration.

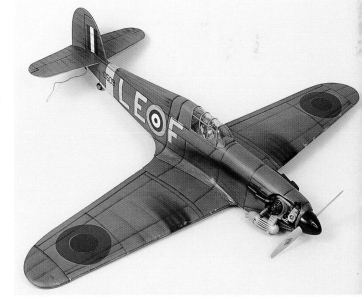

Sports aircraft

Sports aircraft are mostly mid- or low-wing light aircraft types. They are generally less forgiving than high-wing Trainer types, stall more readily, and need more piloting skill. Some model aircraft manufacturers have eased the transition from high-

This semi-scale Hawker Hurricane, from the Cambrian Fun Fighter range, is made from balsa and plywood.

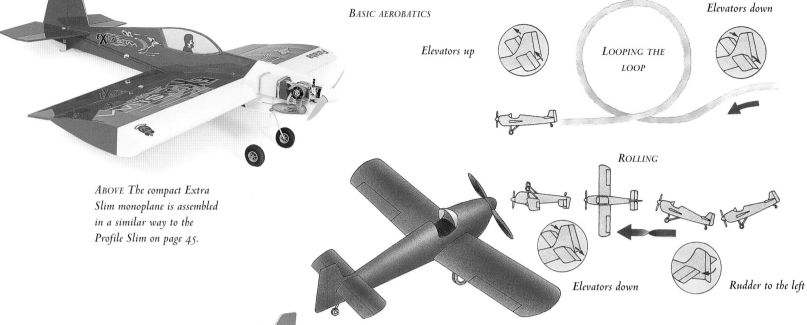

BASIC AEROBATICS

Elevators up

LOOPING THE LOOP

Elevators down

ROLLING

Elevators down

Rudder to the left

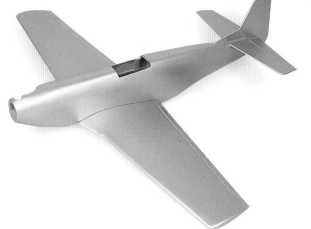

ABOVE The compact Extra Slim monoplane is assembled in a similar way to the Profile Slim on page 45.

LEFT This P-51 Mustang is typical of a more advanced kit model. The airframe is made of wood, with a balsa and tissue covering.

BELOW This flying site near Lockport, Illinois, is perfect for model aircraft, since there are no houses, highways, or big trees nearby.

wing aircraft to low-wing types by making low-wing versions of high-wing Trainers. When you have developed your flying skills with a Trainer, you might like to try the Low Boy 5, which is a low-wing version of the high-wing Hi Boy Trainer. The manufacturers, Precedent, claim that the low-wing type is almost as easy to handle as the Trainer. ARTF Sports kits are also available.

Aerobatics

Sports models with added flaps are known as Aerobatics. These complex models are capable of extraordinary flying feats, and model aerobatic championships are frequently organized. The Slim models also fit into this category. Although they look like a cartoon cutout, these model types fly well, due to their thin fuselage. ARTF Aerobatics kits are also available.

Scale models

These are flying models that reproduce an actual aircraft to scale. If you have several years of modeling and flying experience, building a scale model will be an enjoyable project. Most scale models are big in order to retain their good flying characteristics—four-engined bombers with opening bomb doors are a popular choice. The ultimate in scale models are the professionally made models that feature in movies.

KIT MATERIALS

Traditional kits use balsa and plywood and must be assembled rib by rib, stringer by stringer, just like a real plane. On more modern kits of this type, the wood parts are die-cut and keyed, saving a lot of preparation time. Wood kits are popular as they are lighter than models made of plastic or other modern materials. For the same reason, tissue is still popular as a covering material.

Although ARTF kits come partly assembled or pre-fabricated, they still require some assembly work. Some kits are nearly all wood—usually a mix of balsa and plywood—others are all plastic with molded and sheet parts. More commonly you will find kits made from a variety of plastics: the nose cowl or cockpit area may be made of molded plastic; the wings may be made of foam core (expanded polystyrene), and covered with plastic sheet or veneer. Some older almost-ready-to-fly (ARTF) kits may have a fuselage molded in GRP (fiberglass) although ABS plastics are now more common.

Some simple kits aimed at beginners are made almost entirely out of expanded foam, factory molded and colored. Although they may appear quite crude in kit form, they look surprisingly good when they are assembled and decorated with decals. Airplanes made from expanded foam are easy to assemble and also stand up well to hard knocks in spite of their light weight. The Airdancer featured on page 50 is an example of an aircraft made of expanded foam.

BELOW This model has been made extra strong by the addition of balsa sheeting on the wing roots.

ABOVE In this Precedent Sports Biplane kit, the balsa and plywood parts come die-cut on wooden sheets.

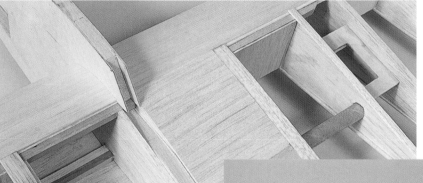

LEFT The ailerons of this Profile Slim model are made separately and then placed in position on the wing.

RIGHT The airframe of this Profile Slim model has been assembled and is now ready to be covered.

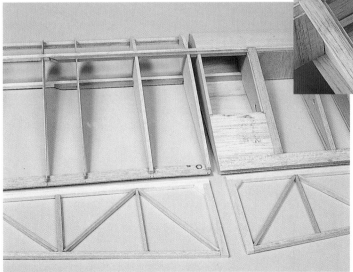

The Kyosho T-33 in Thunderbirds aerobatic team colors is a typical modern jet fighter model. It is made of expanded foam and powered by an electric impeller inside the fuselage.

This cutaway of the T-33 shows how the electric impeller motor fits inside the fuselage to give a jet effect. Note the leads to the electric motor and the servos and linkages (left) for the flight controls.

CONTROLS

The simplest radio-controlled aircraft may have just one channel for controlling the rudder, or two channels for the rudder and engine throttle/speed. Some simpler Trainer models without ailerons have three channels for the rudder, motor, and elevator. However, most models have ailerons and so require four-channel control. On four-channel transmitters the sticks can be moved in two directions, up and down and side to side. As with all other types of radio-controlled models the receiver on the aircraft passes the signal pulses to the appropriate servo to operate its function. A battery supplies the power for this.

If you are operating with other planes, it is important to remember that each aircraft must have its own unique signal. As with radio-controlled cars, boats, and yachts, you can alter the operating frequency by changing the crystal (see page 10).

Although some RTF and other beginner's models, such as the Sky Surfer on page 32, come with all the necessary operating equipment, you will usually have to buy the control equipment separately. Your local hobby dealer will be able to advise you about appropriate equipment. Look out for special offers whereby the kit is sold together with the necessary control equipment.

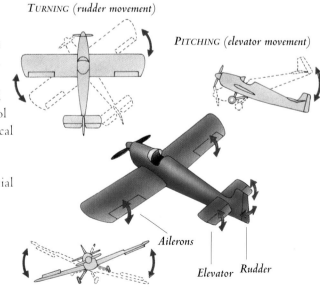

Stringers

Ribs

Propeller

Power unit (IC)

Cockpit

Nose gear

Spinner

The three channels controlling the rudder, aileron, and elevator enable the aircraft to turn, roll, and pitch up and down.

TURNING (rudder movement)

PITCHING (elevator movement)

Ailerons

Elevator Rudder

ROLLING (aileron movement)

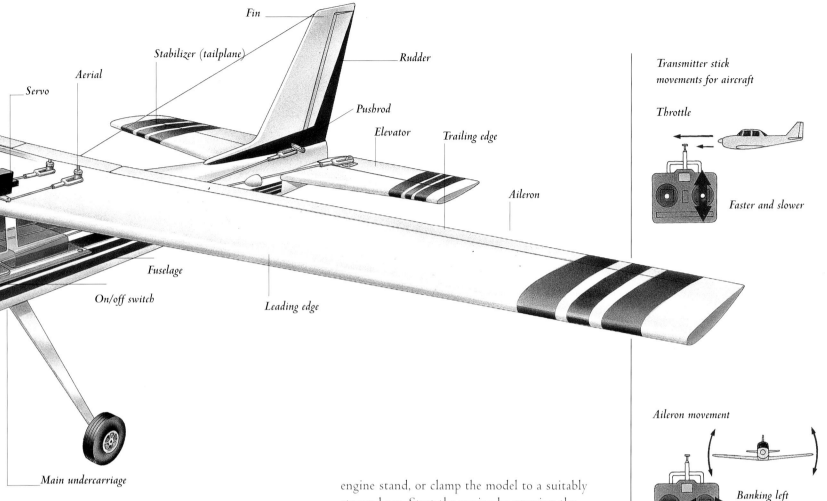

Fin
Stabilizer (tailplane)
Aerial
Servo
Rudder
Pushrod
Elevator
Trailing edge
Aileron
Fuselage
On/off switch
Leading edge
Main undercarriage

Transmitter stick
movements for aircraft

Throttle

Faster and slower

Aileron movement

Banking left
and right

Elevator movement

Up and down
(climbing and
diving)

Rudder movement

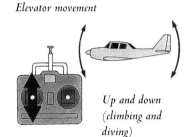

Yawing left
and right

ENGINES

For a long time IC engines were the only source of power for model aircraft, and over the years they have been developed to a high degree of efficiency and quality. Most engines are 2-stroke, although 4-stroke engines are favored for sports, racing, and aerobatic aircraft. The 2-stroke engines are cheaper, however, and are adequate for most models. You will usually have to buy the motor separately—installation details and motor sizes are indicated in the kit and catalog literature. Engine sizes are indicated by their capacity: .10, .15, .20, .25, .40, and .60 cubic inches. Most Trainer airplanes require a .40-cubic-inch engine. You will also need a fuel tank, fuel lines, a propeller, and a small battery to power the glow plug.

It is usual to test and run the engine in for a period on a workbench. You can use a special engine stand, or clamp the model to a suitably strong base. Start the engine by opening the throttle and filling it with fuel. Check the glow plug and flick the propeller to turn the crankshaft and fire the cylinder. This is quite a complex sequence and the flick start of the propeller, in particular, requires some practice. If you find this difficult, you can buy an electric starter that fits over the propeller boss and removes the need for flicking by hand.

In recent years, electric engines have become more popular. These are easy to handle and have none of the exhaust and noise problems associated with IC engines. Although electric engines lack the realism of IC power, no refueling is required, and there is little need for tuning beyond testing the electric circuit—all you need is a supply of charged batteries to keep you flying. Some new kits are now designed to take an electric engine, which is included in the kit together with the appropriate gearbox, and propeller, etc. Other models are designed to take either an electric motor or an IC engine. You can also buy small lightweight engines for micro models. Electric ducted fan motors for jet aircraft that simulate a jet nacelle are also available.

LAUNCHING AIRCRAFT

It is possible to launch some of the smaller, lighter RTF models on your own. Hold the model in your right hand and the transmitter in your left hand, and launch the aircraft into the wind.

Ground takeoff (and landing) is possible if the surface is smooth. Ask a friend to hold the tip of the tailfin until you give the command to let go. The aircraft should run along and take off just like the real thing. However, ground launching needs practice and should never be attempted on the first flight.

LEARNING TO FLY

Although you can run a model car alone in your backyard and quickly pick up expertise, flying a radio-controlled IC-engine aircraft requires great dedication and important skills that are difficult to learn on your own. If you have a simple aircraft, you may be able to fly it with the family in the country, but once you reach the Trainer level, it is essential to join a club. Here, experienced instructors will be on hand to pass on essential flying tips and introduce you to the "buddy" training system. In this system, you fly the aircraft with an instructor who has a duplicated transmitter so that s/he can take over if you go wrong. The club will also be able to advise you of suitable operating sites for IC-engine airplanes. For details of local clubs, ask at your local hobby store or check out current model-airplane magazines.

Simple electric models, such as the Sky Surfer and the Airdancer, are much easier to operate. Although you don't need elaborate training to fly these models, you still need to learn some flying skills, and understand the principles of flying circuits.

Launching aircraft

Before you launch, check your model carefully while holding the model vertically with the nose upward. If your model has an IC engine, check that the fuel tank is full and that the needle valve works correctly both at full throttle and idling speed. If you are launching an electric engine, make sure that the batteries are fully charged and that it runs without hesitation. Test all control surfaces to check that they work without restriction and that they respond to commands from your transmitter. Finally check that the propeller is free to revolve.

It is helpful (and safer) to ask a friend to help with hand-launching. Ask your friend to hold the aircraft while you control the launch. Stand slightly to one side, give the command to let go, and then launch the aircraft into the wind.

BASIC CIRCUIT

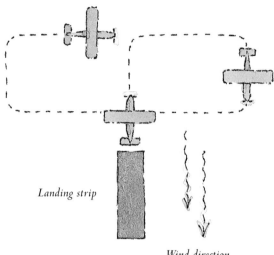

Landing strip

Wind direction

Determine the exact direction of the wind—you can use the ribbon pendant on your aerial as a windsock. The aircraft will be launched in this direction.

Check that the area is clear and note the upwind and downwind skylines as you will have to judge the turning points by eye.

INSURANCE

Third-party insurance is essential for all model aircraft, except perhaps for the very simplest beginner's models. A large aircraft with a wingspan of 60in (150cm) is heavy and can cause serious injury or damage to property. Consult your local club, hobby dealer, or current model-aircraft magazines for more information.

MAINTENANCE

As with all models, constant maintenance is essential. During flying sessions you can expect minor damage, such as tears in the covering, and these should be repaired before you operate the aircraft again. Prevent accidental damage by having a good carrying case for your model(s). Also useful is a portable tool box containing all the tools you need for on-the-spot repairs and general maintenance.

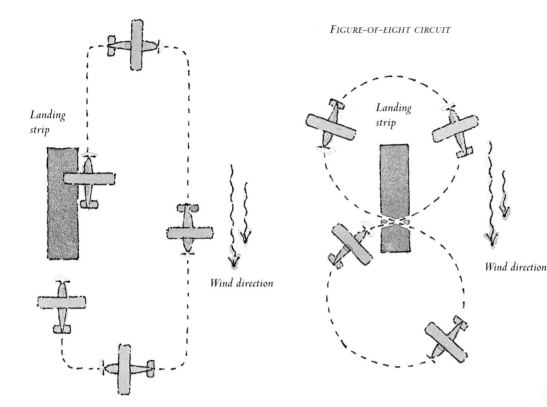

FIGURE-OF-EIGHT CIRCUIT

Landing strip

Landing strip

Wind direction

Wind direction

Flying your model in regular circuits will help to improve your flying skills. You can try a variety of circuits, ranging from a simple test circuit to a more challenging path around a wider area.

SIMPLE OUT-AND-BACK CIRCUIT

Pilot's view of his model being hand-launched by his assistant at a model show.

DOS AND DON'TS OF MODEL-AIRCRAFT FLYING

- Only fly on approved sites, unless you have a small, simple, electric beginner's model.
- Never fly near bystanders or animals.
- Do not fly near buildings of any kind.
- Do not fly near pylons or power lines.
- Do not fly by (or across) railroad tracks, highways, sidewalks, or public paths.
- Do not fly near airports, real aircraft landing strips, or under aircraft flight paths.
- Do not fly in strong winds or rain.
- Even on an approved flying site, do not operate unless you have a clear space of at least half a mile (1km) around you.
- Do not fly unless you can keep the aircraft in sight for the entire flight—never direct the aircraft beyond distant trees or hilltops.
- Always take off and land into the wind—put a ribbon on your aerial to act as a windsock.
- Always check your frequency with other flyers present, changing crystals as necessary.
- Take care when launching and always brief any helpers on your "flight plan."
- Take care when refueling and testing the engine.
- Never touch the motor immediately after flight. It may be very hot.

The stylish Robbe Airdancer is an impressive model and is easy to build and operate. With its wingspan of 4ft 9in (145cm), the Airdancer is a great aircraft for demonstrating your skill at turns, loops, and climbs.

A STYLISH AIRCRAFT

- **Easy to launch and fly**
- **Easy to assemble**
- **No painting required**

Making and flying a radio-controlled model aircraft is an ambitious project for a beginner. If you are not careful to choose the right model according to your skills and experience, your attempts at flying your model may end in disappointment. However, in recent years, some model manufacturers, such as Robbe, have started to make a number of kits designed for the inexperienced pilot or the newcomer to radio-controlled models. These beginner kits are also suitable for young modelers, since they are safe and easy to assemble, although adult supervision is useful.

The Airdancer is one of the biggest and most impressive of these beginner kits, with a wingspan of 4ft 9in (145cm). The parts are all molded in colored styrofoam, bright yellow in this case, and no painting is required, since the kit comes with self-adhesive decals. All the parts you need to build the model and a set of clear instructions are supplied with the kit. The kit is easy to assemble, as all the parts either glue together or are held by self-adhesive tape. Apart from a screwdriver and a craft knife, you need no other tools.

TOOLS YOU WILL NEED
Small screwdriver
Craft knife

OTHER REQUIREMENTS
Control Set 500
One ni-cad 7.2V battery pack
Batteries for transmitter

Controls
There is a specially produced companion set, Control Set 500, which contains all the control equipment you need to get ▶

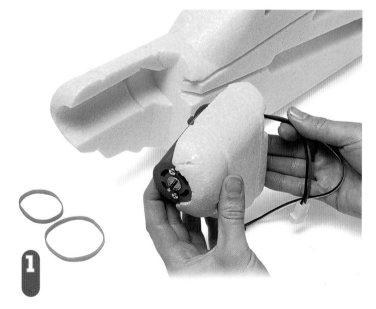

1 *The fuselage comes as a solid molding with slots for installing the controls and a separate engine nacelle. Start by identifying the electric motor and fit it into the molded recess located within the nacelle.*

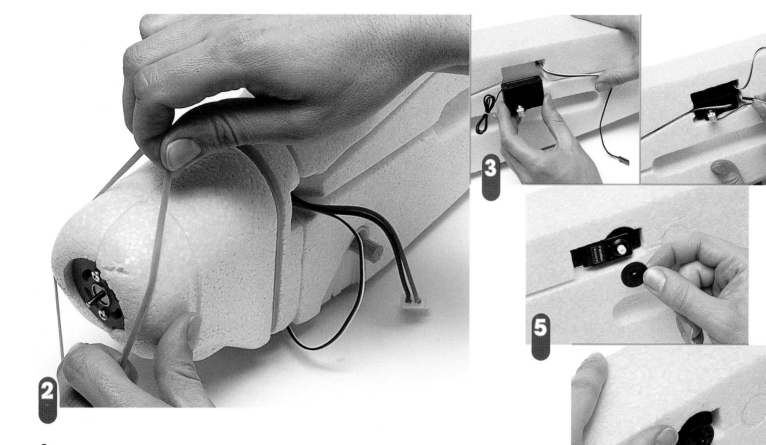

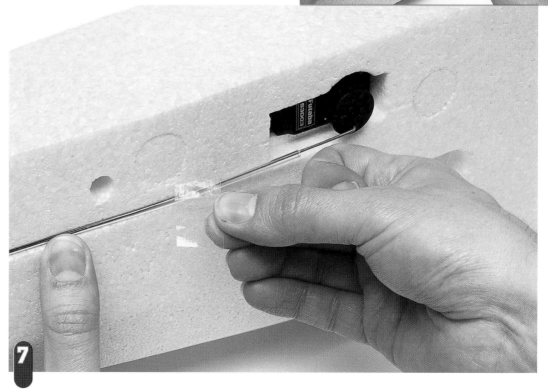

2 *Ensure the motor leads are clear, and push them into the slot in the fuselage side. Push the engine nacelle into position on the front of the fuselage and secure it using the rubber bands provided in the kit.*

3 *Identify the electronic speed controller from the control set and then force-fit it into the recess provided in the side of the fuselage. Make sure that the wire connections hang free and are not jammed inside.*

4 *Wire up the connections and take the leads forward to the electric motor by inserting them into the narrow slot in the fuselage side. Another wire to the rear joins to the receiver unit, and this is also slotted into the fuselage.*

5 *Insert the servo controlling the rudder into the appropriate recess in the fuselage. Take the top disk off the servo, so that it can be connected to the wire pushrod which will move the rudder.*

6 *With the rudder operating rod attached, screw the disk back onto the spindle of the servo.*

7 *Fit the rudder operating pushrod deep into the narrow slot provided for it—use strips of adhesive tape to ensure the rod does not inadvertently fall out of the slot, though the tape should not normally touch the rod.*

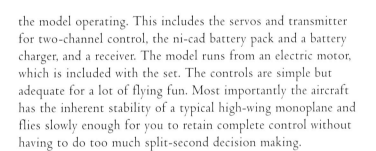

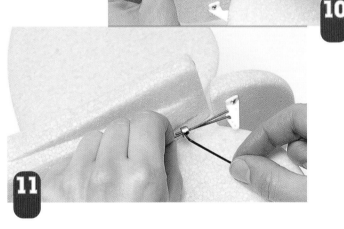

the model operating. This includes the servos and transmitter for two-channel control, the ni-cad battery pack and a battery charger, and a receiver. The model runs from an electric motor, which is included with the set. The controls are simple but adequate for a lot of flying fun. Most importantly the aircraft has the inherent stability of a typical high-wing monoplane and flies slowly enough for you to retain complete control without having to do too much split-second decision making.

Flying the Airdancer

To fly the Airdancer, simply switch on the motor and transmitter, and launch the airplane from the shoulder with a gentle forward push. The Airdancer will climb away and you can then direct the plane from the transmitter by adjusting the rudder and controlling the speed of the motor. With a little practice you can achieve turns, loops, climbs, and even stall turns. When you want to land, switch off the motor and the aircraft glides down to a flat landing in the grass. You should not attempt a wheeled landing with the Airdancer, since the undercarriage is cosmetic and will not support the weight. Because the model is so simple, it is ideal for a solo flyer, although it is helpful to have two people—one to launch and one to use the transmitter—for early test flights. The model is robust and tough and as long as you do not fly it into trees or onto hard landing surfaces, it should have a long life.

8 *The stabiliser is another styrofoam molding. You can now insert it into the slot provided for it at the rear of the fuselage.*

9 *Screw the plastic horn that takes the rudder operating pushrod onto the molded rudder—it is important to check carefully that it is square and secure, as the rudder will see a lot of movement.*

10 *Adhesive tape is provided in the kit to make the rudder hinges. This is an easy task, but be sure to keep the rudder perfectly aligned and upright while the tape hinges are applied.*

11 *With the rudder in place on the tail fin you can now join the rudder operating rod to the horn and make the connection.*

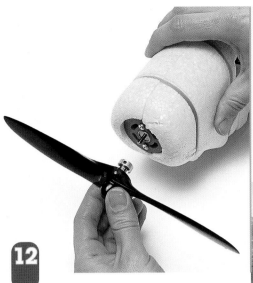

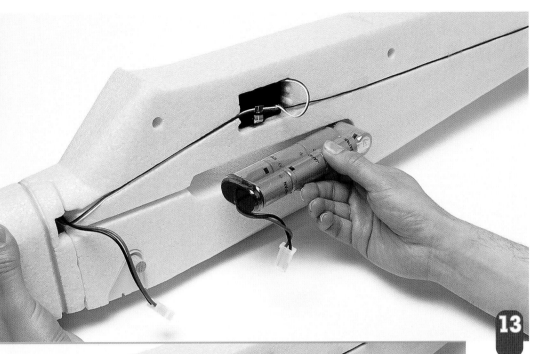

12 Screw the molded propeller to the motor shaft and check that it revolves freely and accurately.

13 Fit the battery pack (which must be charged before every flying or test session) into the appropriate slot in the fuselage side and connect the receiver and speed controller. At this stage you can switch on the transmitter to check that the rudder and motor control functions work correctly.

14 Insert the wooden rods provided with the kit to support the wings on the fuselage top and the undercarriage at the bottom. In both cases, use rubber bands to secure the wings and wheels in place, so that they will knock off easily in the event of a heavy landing.

15 Assemble the wheels and tires and locate the undercarriage struts and check them for fit.

16 Secure the undercarriage to the lower wooden rods with elastic bands. Note the use of yellow adhesive tape under the wheel struts that prevents them from cutting into the styrofoam fuselage.

17 Now assemble the three-part wing—two outer panels and a center. There are no wing controls on this model, making it a simple model to operate.

18 *Join the wings together using yellow adhesive tape to strengthen the styrofoam moldings and protect them from being worn down by the elastic bands over the wings.*

19 *As in all traditional model aircraft, the wings are held in place by attaching them via rubber bands to the rods through the fuselage.*

20 *Construction is now complete and your final task is to affix the decorative markings. Cut out the decals, position them carefully, and rub them down with a damp, slightly soapy cloth to eliminate air bubbles.*

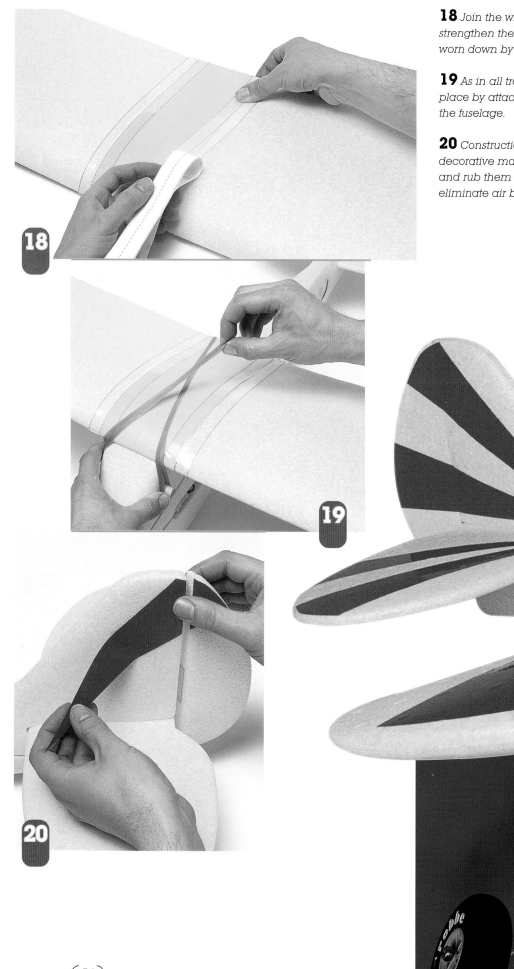

The Robbe Airdancer model is now complete with all markings applied, including the cockpit windows. Note that the trim strips conceal the fuselage slots holding the control components, giving the model a remarkably realistic appearance.

FLYING THE MODEL
Although the Airdancer is one of the simplest aircraft kits available, it offers a great introduction to "real" radio-controlled flying. It is easy enough for a young enthusiast to build and operate, because there is a control set (Control Set 500) specifically designed for this model.

SEE ALSO
Control equipment, p.19
Flying an aircraft, pp.46–49
Introduction to aircraft, pp.38–39

Invariably graceful and elegant, radio-controlled gliders are a popular choice among modelers. They are also easier to build and often less expensive than other types of radio-controlled models. As most gliders do not need an engine, there are fewer restrictions on where you can fly.

INTRODUCTION TO GLIDERS

Gliders use thermals and air currents for flight. They are dependent on their launch momentum to get airborne and find the currents or thermals to keep them up. Thermals are rising currents of warm air, which vary according to the temperature of the ground below. The process of circling to find the next thermal is known as "thermal hunting." With a combination of pilot skill and good weather conditions, the glider can be flown for 20 minutes or more. If the glider is high enough and the thermals are strong enough, interesting aerobatics, such as loops and rolls, are also possible. As on a powered model, the ailerons, elevator, and rudder are controlled by radio.

Powered gliders are a recent innovation, enabling extra height to be gained at launch, so that they can reach high thermals over 500ft (150m), and stay up longer. These are similar to real gliders that have low-powered motors to get them up really high. Powered model gliders have a small electric motor in the streamlined nose that drives a small folding propeller. The battery pack is also on board. The motor is started and the glider is launched (usually from the shoulder) just like a conventional powered aircraft. However, when the glider reaches the desired height, a servo in the control system switches the engine off. The propeller blades then fold flat over the nose and unpowered flight takes over.

Many glider kits today can be built in powered or unpowered forms. The motor and

RIGHT A variety of basic flight operations and maneuvers are possible with radio-controlled gliders. You can launch by hand or use a tow or bungee to reach operating height.

ABOVE Thermals are rising currents of warm air, caused by the uneven heating of the ground by the sun. Gliders can use thermals to gain height, and skilled pilots can prolong flight by locating strong thermals.

4 LANDING Reduce speed; use the airbrake if fitted

5 TURN Position the aircraft so that you can land it into the wind

propeller are included in some powered glider kits, but in others you can choose the type of motor installation yourself. Several different sizes of motors are available for powered gliders, depending on the size of your model. Typical motors are for a 7.2V or 8.4V operation, but other options are available. The Graupner Speed range, for example, offers glider motors in 6V, 7.2V "mini" (slimline), 7.2V standard, and 8.4V forms.

Some sophisticated options can be added to the larger sailplane models. For example, you can add different wing airfoil sections for extra performance. Wings suited to different flying methods, such as slope soaring or riding thermals, are available, so that you can choose the kind of wings you use according to the operating conditions.

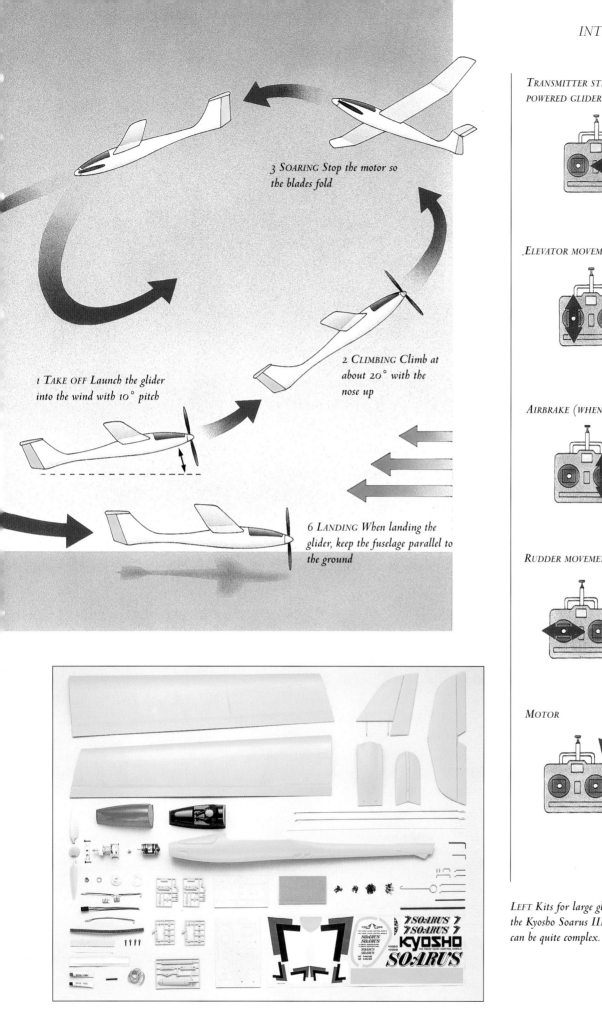

3 SOARING Stop the motor so the blades fold

1 TAKE OFF Launch the glider into the wind with 10° pitch

2 CLIMBING Climb at about 20° with the nose up

6 LANDING When landing the glider, keep the fuselage parallel to the ground

TRANSMITTER STICK MOVEMENTS FOR A POWERED GLIDER

Banking left and right

ELEVATOR MOVEMENT

Up and down (climb and dive

AIRBRAKE (WHEN FITTED)

Up and down

RUDDER MOVEMENT

Yawing left and right

MOTOR

Function switch off and on

LEFT *Kits for large gliders, such as the Kyosho Soarus III, shown here, can be quite complex.*

CONTROLS

Basic gliders use two channels, which control the rudder and elevator. The minimum requirement for a powered glider is three channels—the third channel controls the motor via an electronic speed controller and switch. More advanced gliders require a fourth channel to control the ailerons. You may also find gliders with airbrakes that need a fifth channel. All radio-controlled gliders have a receiver and the necessary battery power on board. As with other types of radio-controlled models, start off with a simple glider. Once you have mastered the basic construction and flying skills, you will have the confidence and experience to try a more complicated model with advanced features, such as airbrakes.

LAUNCHING A GLIDER

Basic models are launched using a throw (for small gliders) or a winched or towed launch on a cord. This is designed to drop free when the model reaches a certain height, usually 500ft (150m) maximum. For extra momentum, you can launch a glider using a rubberized cord and hook. Tow ropes and winches of various kinds are available from your local hobby store. However, hand-launching from the shoulder is also possible for test flights and in very light airs.

A popular alternative to ground launching is slope soaring. You will need a hill face with the wind blowing onto it, as the glider is launched downhill into the wind. With this type of launch, you can expect an exciting flight, especially if

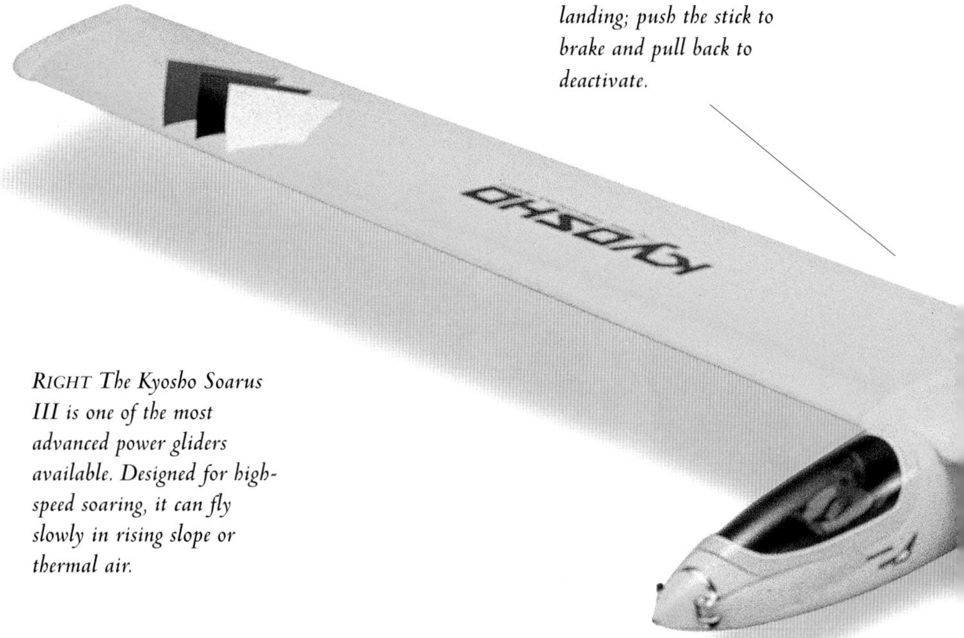

AIRBRAKE, WHEN FITTED (POWERED GLIDERS ONLY) Used to reduce speed when landing; push the stick to brake and pull back to deactivate.

RIGHT The Kyosho Soarus III is one of the most advanced power gliders available. Designed for high-speed soaring, it can fly slowly in rising slope or thermal air.

PROPELLER ACTIVATION (POWERED GLIDERS) Allows glider to take off and climb; cut the power when soaring altitude is reached.

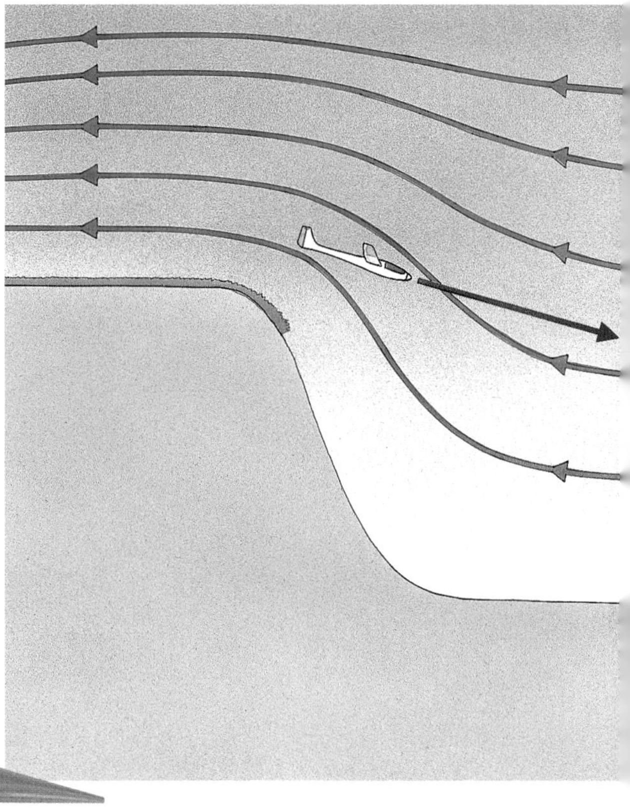

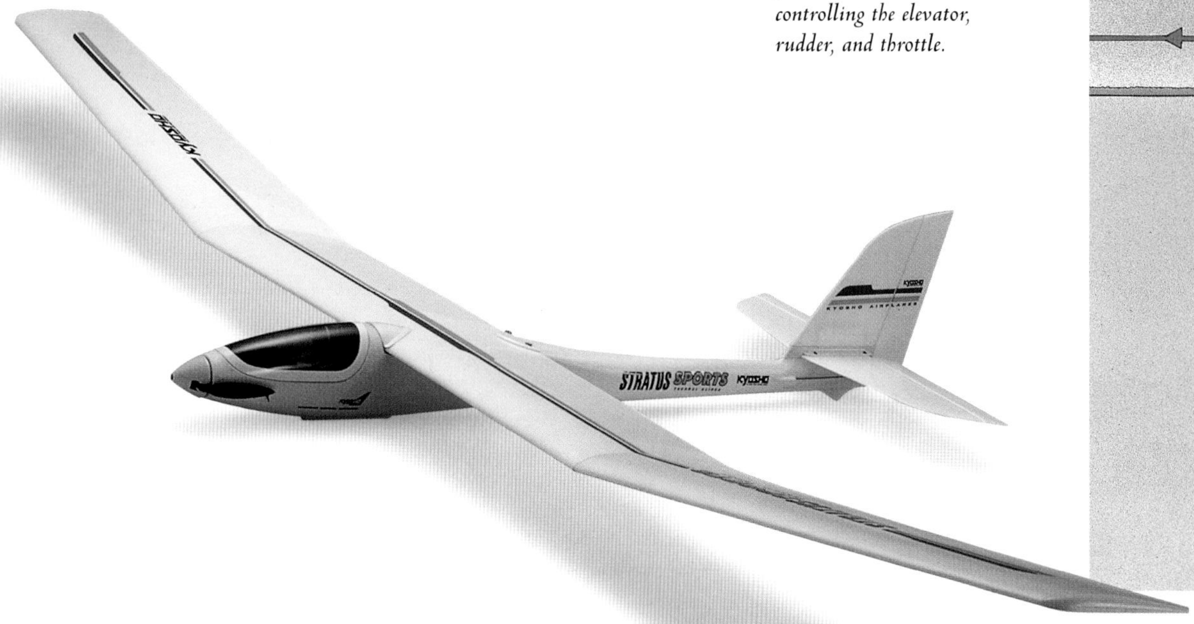

BELOW The large Kyosho Soarus, with its wingspan of 72in (182cm), is an advanced glider with three channels controlling the elevator, rudder, and throttle.

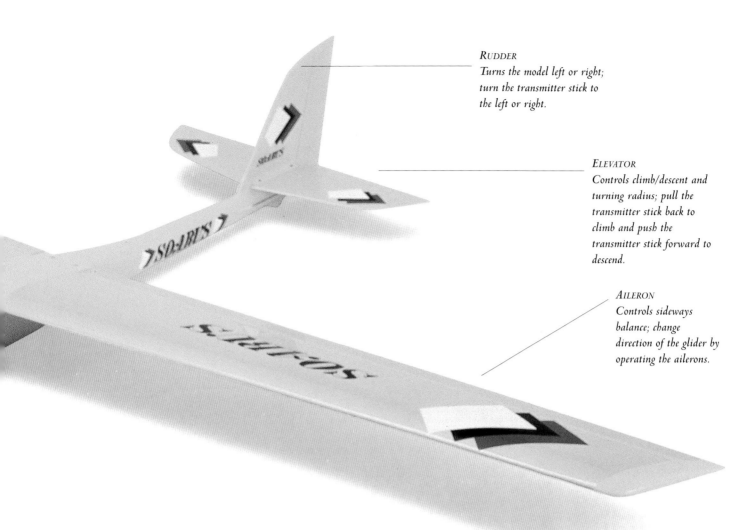

RUDDER
Turns the model left or right; turn the transmitter stick to the left or right.

ELEVATOR
Controls climb/descent and turning radius; pull the transmitter stick back to climb and push the transmitter stick forward to descend.

AILERON
Controls sideways balance; change direction of the glider by operating the ailerons.

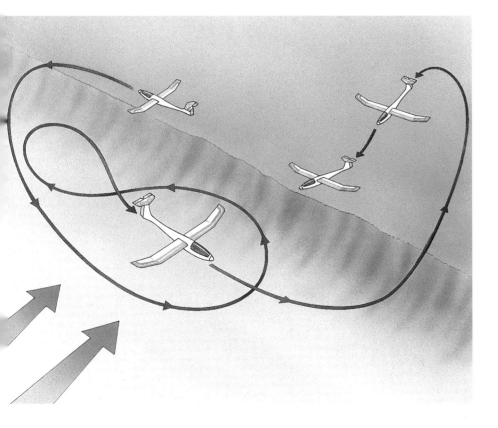

there is a strong breeze. If you are launching a powered glider, be careful not to climb it at too steep an angle, as it may suddenly stall, spin, or crash, if there is a drop in speed.

FLYING A GLIDER

The length of time you can keep your glider in the air is determined partly by the wind conditions and partly by the skill of the pilot. Flying a glider successfully for an extended period will depend on your ability to recognize thermals and updrafts and your ability to launch your model into them. A skilled pilot can keep a glider up for 20 minutes or more, but don't be disappointed if your first flight lasts only a few minutes, since you will improve with practice.

LEFT Winds moving toward a hillside become natural updrafts, making a slope an ideal place to find ascending currents of air.

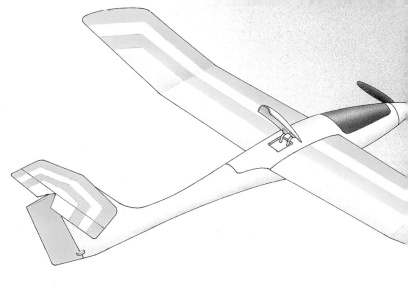

You can also buy radio-controlled model gliders that are built to scale. This superb model is a scale replica of the elegant Minimoa glider built in the 1930s.

FLYING A POWERED GLIDER

Before you launch, check the motor carefully. Make sure that the freshly charged ni-cad battery is correctly wired and mounted in the fuselage. Check the transmitter controls and then switch on the motor (the method will depend on the model) and launch the glider. To make full use of the limited life of the battery, start the motor at the last minute when ready to launch. Climb at a gentle angle to avoid stalling. When you reach the desired height, switch off the motor. You can now fly the model just like an unpowered glider.

SAFETY AND RULES

All the safety rules and operating tips for powered flight (see page 49) apply equally to gliders, although most operating restrictions on flying sites do not apply to gliders.

Once you have some flying experience, you can try some aerobatic maneuvers. Always practice into the wind, since aerobatics down or crossing the wind are difficult.

The elegant Graupner Elektro Junior, shown here in flight with its engine running, is a sophisticated powered glider.

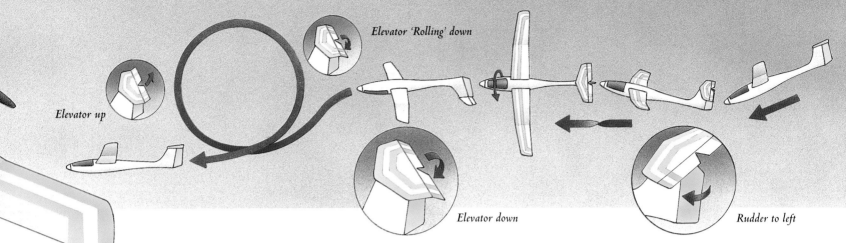

Elevator 'Rolling' down

Elevator up

Elevator down

Rudder to left

LOOPING THE LOOP

Apply the down elevator to lower the nose and increase airspeed. When sufficient speed is obtained, apply the up elevator to raise the nose. Keep the elevator back and the glider will continue up and over into a loop. When the loop is complete, return to straight and level flight.

ROLLING

Apply the down elevator to lower the nose and gain airspeed. Now bring the nose up and move the rudder in the direction you want to roll. When the glider is upside down, add down elevator to keep the nose from falling out of the roll. When the roll is almost complete, move the rudder slightly in the opposite direction to bring the fuselage back to the horizontal.

Ideal for newcomers to powered gliders, the basic Robbe Quincy has two-channel control and is easy to assemble. The wingspan is 64in (160cm).

Radio-controlled gliders are a popular choice because they are less complex than powered aircraft. A typical example of a well-designed kit suitable for more advanced modelers is the Alt Stream glider manufactured by Tamiya.

AN IMPRESSIVE GLIDER

• All the features of a real glider
• Drilling and cutting required
• Six-channel radio control

The Alt Stream kit makes up into a very big glider with a wingspan of 80in (2000mm) and a length of 41.2in (1045mm). It comes in a very big, sturdy box with a carrying handle so that it can be used to transport the model to the flying field. As is usually the case with flying models, the wings are detachable and fit in the box next to the fuselage.

Skills required

Although the Alt Stream is one of the easier large radio-controlled glider kits, it is definitely not a quick project and you should not attempt this project if you are a complete beginner. Some drilling and cutting of wood formers is required, although the main structure is made of ABS vac-form plastic or machined styrofoam (in the case of the wings) for lightness. Some construction tasks are particularly difficult and time-consuming. The control rods and associated bell cranks to operate the rudder, elevator, and ailerons are long because of the size of the glider and are especially tricky to fit. It is, however, a very satisfying project.

TOOLS YOU WILL NEED
Emery board
Craft knife
Selection of large and small
 screwdrivers
Mini-drill

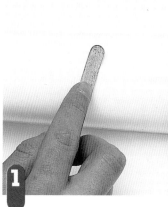 

1 *Before you start assembling your model, check the fuselage molding for small mold marks and raised ridges of flash which may need gently rubbing down with an ordinary emery board. Some similar trimming (with a craft knife) may be needed on the wings.*

2 *The tail needs careful work, in particular at the top where the stabiliser is fitted. File it down flat so that the stabilizer sits perfectly horizontal on it without tilting to the left or right.*

3 *Hinge*

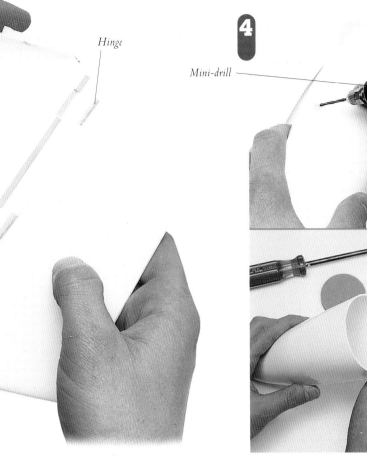

Mini-drill

Hinge

3 Now glue the small plastic plates that take the hinges for the moving control surfaces in place using epoxy cement. Here the rudder is being offered up to the tail fin to ensure that the hinges and the mounting plates are all perfectly in line. This is an extremely important step as all control surfaces must move freely.

4 Vac-formed parts, unlike plastic molded kits, do not usually have the screw holes provided. In the Alt Stream kit all the screw holes of the vac-formed parts have to be drilled out. However, the drill positions are clearly marked by molded dimples to make the task easier. Here, a mini-drill is being used to drill out the holes that take the locating rods for the wings. You will also need to drill similar holes on top of the tail fin so that the stabilizer can be attached.

5 Weight the model using the metal balls provided in the kit. Push the weights into the nose cone with the handle of the screwdriver, then hold in place with the pink plastic bung supplied with the kit.

6 Identify the interior supports, or formers, which are all made of plywood. These are used to support the wings, reinforce the fuselage, and act as mounting points for the servos and the receiver and battery pack. Carefully nick out the accurately shaped parts from the die-cut sheets.

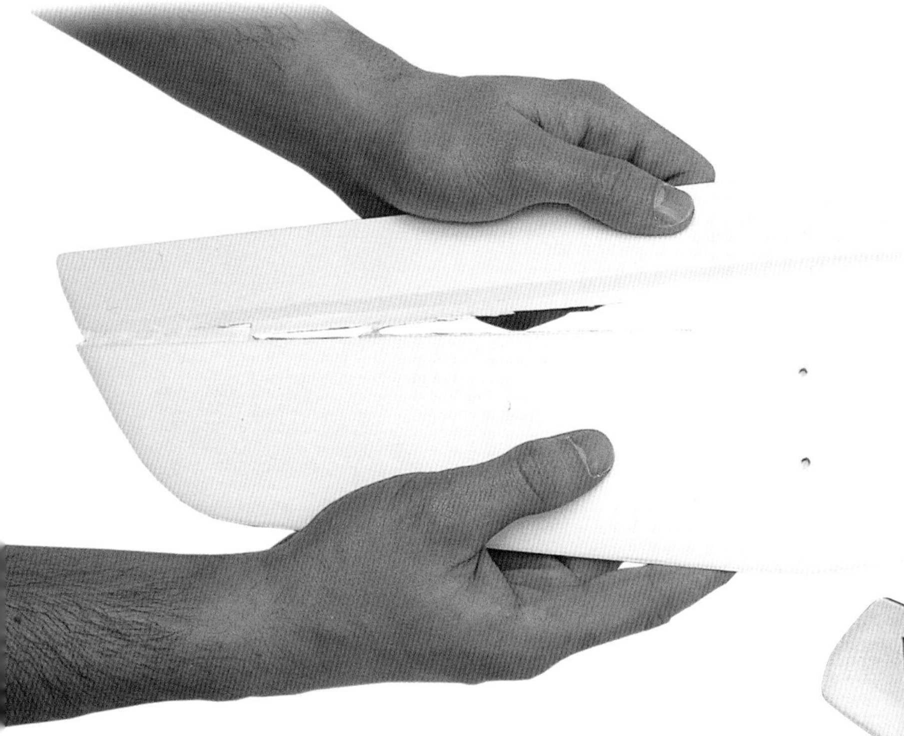

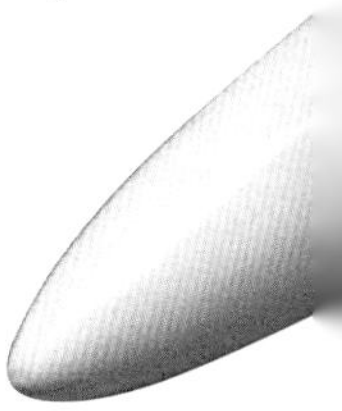

7 The moving control surfaces come already attached in position. However, they must be removed, cleaned up, and fitted with hinges and horns before they can be fitted properly in position. To do this, you will need to remove the elevators from the stabilizer, as shown here.

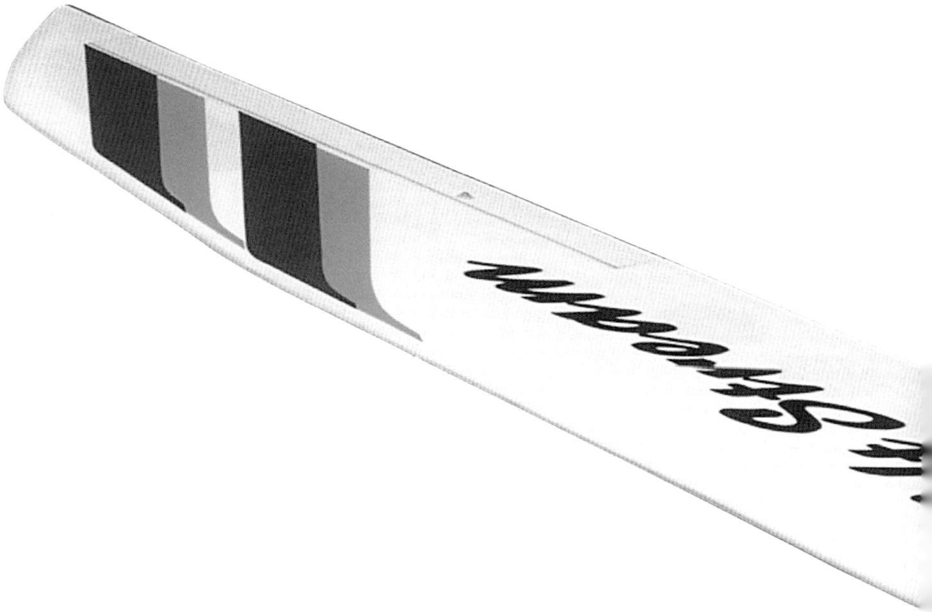

8 Now cut out the decals and fix them carefully to the glider. Here, decals are being affixed to the wing.

9 Because of the size and weight of the wings, they are held in place over rods inserted through the fuselage. The rods fit into sockets in the wing roots, but the wings remain loose so that they are readily "knocked off" in the event of a heavy landing.

A view of the control equipment
installed in the fuselage.

The completed model has an
impressive bird-like appearance. Due
to its configuration and ability to
exploit thermals for duration flights,
it performs extremely well in
optimum flying conditions. The
standard Alt Stream model is
launched in the conventional way by
hand or with a tow-line. However,
power launching is also possible
using an electric motor that is
available as an extra option for this
model. A nose with folding propeller
is produced containing the motor
and substitutes for the plain
nose supplied.

FLYING THE MODEL

*This is a difficult model, suitable for
experienced kit builders. Flying the
Alt Stream is like flying a real glider,
because you can control the ailerons,
rudder, elevator (and motor speed on
the power version). All the control
equipment is mounted on formers in
the fuselage beneath the wing
mounting, with pushrods from the
servos to horns on the moving control
surfaces. This model uses six-channel
radio control in order to simulate the
full range of control functions that
are possible on a real glider.*

SEE ALSO

Control equipment, pp.58–59
Flying a glider, pp.60–61
Introduction to gliders, pp.56–57
Thermals, pp.58–59

The Stratus Sports soarer is an exciting example of how recent advances in model technology have improved the performance of model gliders. It is fitted with an electric motor which enables it to climb quickly in search of thermals, just like a real power glider!

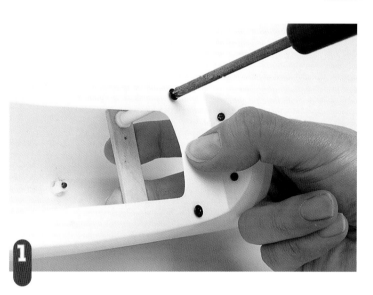

A POWERED SOARER

• Three-channel radio control
• An electric motor for longer flights
• Modeling experience required

Nowadays many large radio-controlled gliders, such as the Stratus Sports soarer, are equipped with an electric motor and a propeller in the nose. These enable the craft to climb quickly to find thermals so that it can enjoy duration flights. The powered soarer is turned back into a glider by cutting off the motor when the propeller blades automatically fold in tight to the nose.

Manufactured by Kyosho, the Stratus Sports is a big aircraft with a wingspan of 76.8in (1920mm) and a length of 40.6in (1030mm). It has a simplified control system, however, as only the rudder, elevator, and speed are working functions and there are no ailerons on the wings to worry about. For this reason it can be operated with three-channel radio control. As is usual, the appropriate control equipment (Futaba is recommended) must be purchased separately; your local hobby store should stock suitable items.

TOOLS YOU WILL NEED

Emery board
Craft knife
Selection of large and small screwdrivers
Mini-drill

OTHER REQUIREMENTS

Ni-cad battery of 8.4V rated at 1.7 amp/hour

Assembly

The Stratus Sports is billed as suitable for newcomers and intermediate modelers, but it is not really suited to complete beginners. Although the wings and fuselage are fully shaped and complete, assembling the kit requires some preliminary work.

1 *First you will need to do considerable preparation work on the fuselage before you can start assembling the model. Several items must be installed in the fuselage, including a pre-cut plywood tray which is the mounting for two servos. However, an additional section must be cut from this tray so that the servos fit properly. It is a good idea to mount the servos on blocks of wood to give the servos sufficient clearance from the bottom of the fuselage. The kit comes with "star-shaped" horns on the servos but to make them fit, they need to be modified to the "single horn" type by cutting away the other three horns forming the star shape. The holes in the horns will not be big enough to take the pushrod linkages, so you will need to open them out by careful drilling before installing. Take great care doing this to avoid damaging the servos.*

2 *Before putting the trays and control items in the fuselage, fit the electric motor into the nose. This is a tricky task, because two screw holes need to be lined up from the motor to a front plate with the fuselage front sandwiched between them. To make the motor fit correctly, it is necessary to discard the two plastic washers supplied with the kit.*

3 *No means of supporting the electric motor are provided with the kit and there are no suggestions in the instructions. However, a good way to keep the motor horizontal is to pack some sponge rubber, cut to size, under the motor. Next attach the two leads provided in the kit to the motor; these will be connected to the speed controller when it is installed later on.*

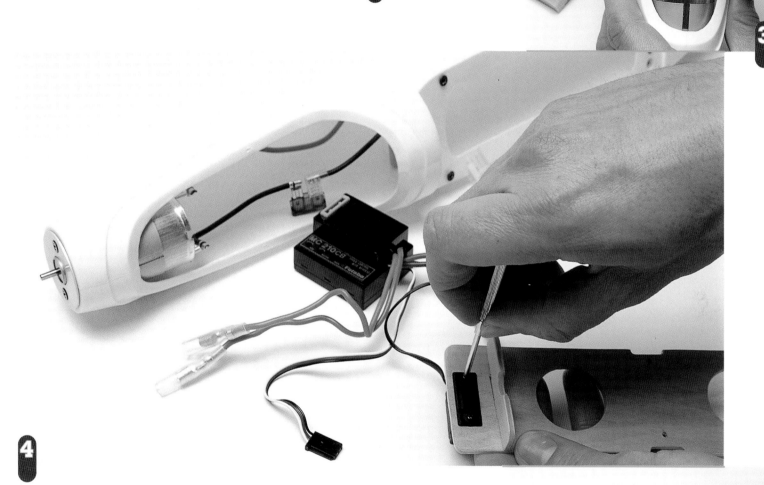

4 *Now connect the electronic speed controller to the motor. Join the red lead to the motor's positive terminal and the black lead to the motor's negative terminal. You also need to wire the speed controller to a switch screwed to the plywood battery tray, as shown here; the plywood tray will be installed later.*

5 *Next to be fitted is the receiver. Connect this to the servos and the speed controller and then connect the aerial line to the receiver. When all the connections are complete, wrap the receiver in the sponge rubber provided in the kit and hold it in place with a rubber band. When the assembly is completed, it is best to place both the speed controller and the receiver in the forward fuselage. This is at variance with the instructions, which advise you to position them above the battery tray. However, you will find that there is insufficient clearance to do this.*

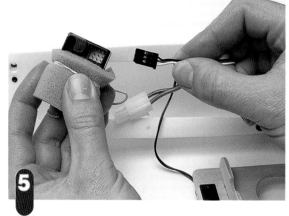

6 *Thread the aerial wire from the receiver through the plastic tube provided with the kit, and fit the tube inside the fuselage, so that the wire extends behind the aircraft through a hole previously drilled out under the tail.*

7 *Attach the ni-cad battery to its own plywood tray using the strip of Velcro supplied in the kit. This will help to prevent it moving about in flight. Now connect the battery to the remaining connection on the speed controller. Pass the pushrods that operate the rudder down inside the fuselage. To complete the control installation, attach the ends to the servo arm, as shown here.*

8 *Now assemble the tail simply by pegging it into place on top of the fuselage. (The necessary holes will have been drilled during the initial preparation stage.) The pegs also hold the stabilizer in place and allow the assembly to be removed when necessary.*

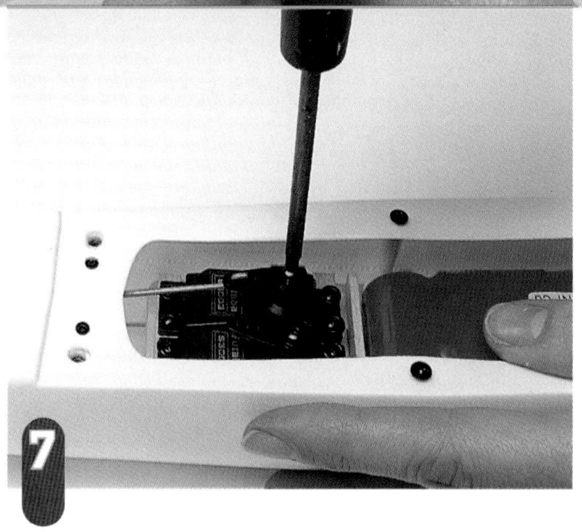

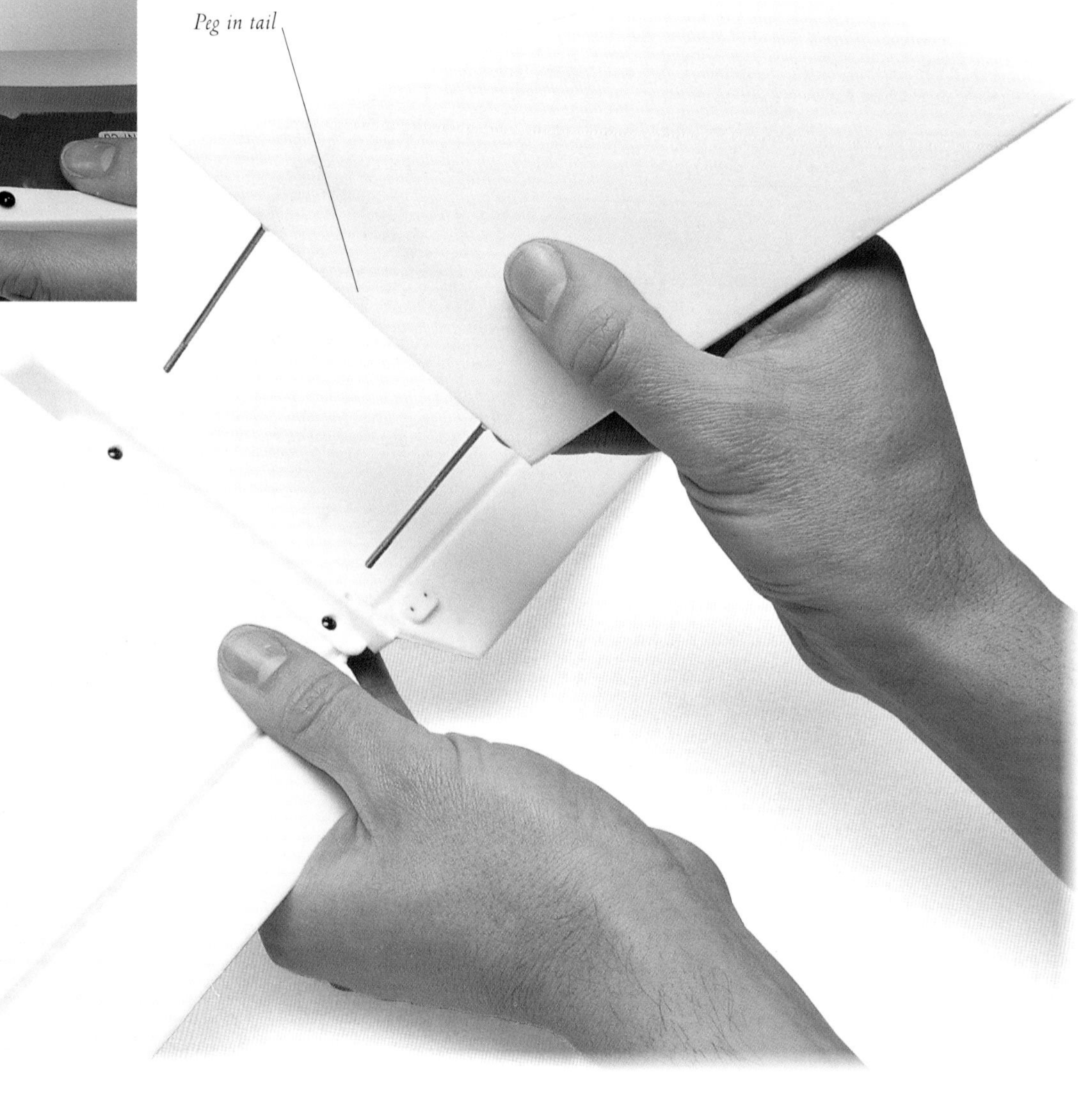

Peg in tail

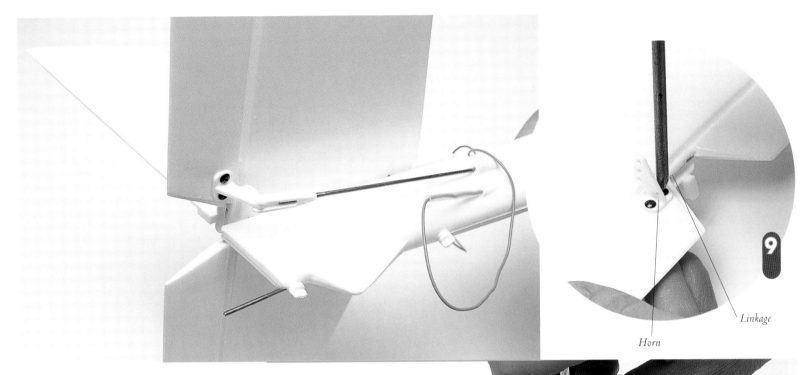

Linkage

Horn

9

9 *Screw the horns connecting the push-rods to the rudder and elevator into position, in this case on the elevator. The pushrod attached to the horn on the elevator passes through the hole in the fuselage to connect to the servo shown in stage 7. Attach the aerial to the receiver inside the fuselage and trail it out of the lower hole.*

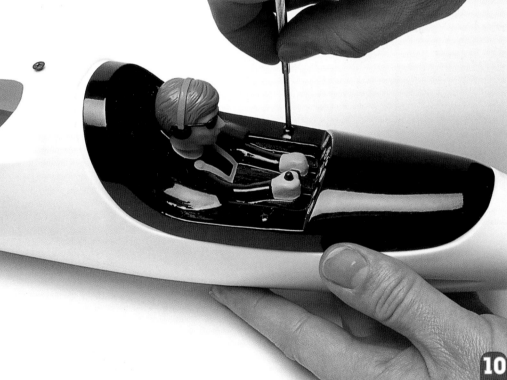

10 *To complete the fuselage construction, assemble the plastic cockpit insert and the separate cockpit canopy. First cut them out of the pre-formed plastic sheets and then paint the cockpit insert. You will also need to paint and attach the pilot's head, which is a separate molding. Position the "instrument panel" sticker in front of the pilot. Now screw this assembly into place so that it can be removed when access to the equipment inside the fuselage is necessary. Attach the cockpit canopy with double-sided adhesive tape, so that it too can be removed.*

10

11 *Now assemble the propeller and spinner. The two propeller blades are hinged, so that they fold back to the fuselage sides when the motor is turned off in flight. Use the Allen key supplied with the kit to tighten up the cap screw in the front end of the spinner.*

11

12 *The wings come in three pieces (center, outer port, and outer starboard) and are made of foam core with a thin plastic skin. The outer sections have pronounced dihedral angles, and the center section has wooden spars molded in. To assemble the wings, attach each outer wing section to the center by the spars, also using the plywood inserts that plug into slots on the inner faces of each section. The inserts have the correct dihedral angle cut into their shape. Epoxy cement is recommended for all the wing joints to ensure a strong wing structure.*

13 *To attach the wing, insert the metal rod supplied with the kit into the center leading edge of the wing. This fits into a slot cut into the aperture that was made during the earlier stages of construction.*

14 *Complete the wing attachment by screwing down through a plywood reinforcement plate into two holes drilled at the rear of the wing aperture cut out in the first stage of construction. This is quite a difficult task, as the vac-form plastic tends to compress under the pressure of the screwdriver. Cut a shaped plastic fairing from the vac-form sheet provided to cover the leading edge wing joint. You can attach this with double-sided adhesive tape so that you can easily remove the wing to get access to the control gear.*

15 *Finally, decorate the completed model with the self-adhesive decals supplied with the kit. Cut them out carefully using small scissors.*

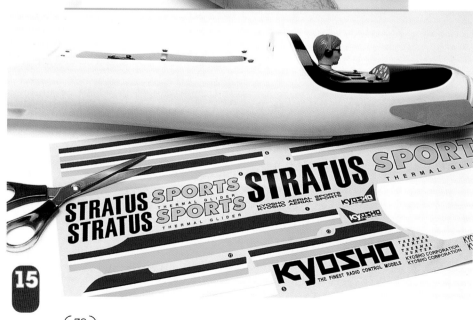

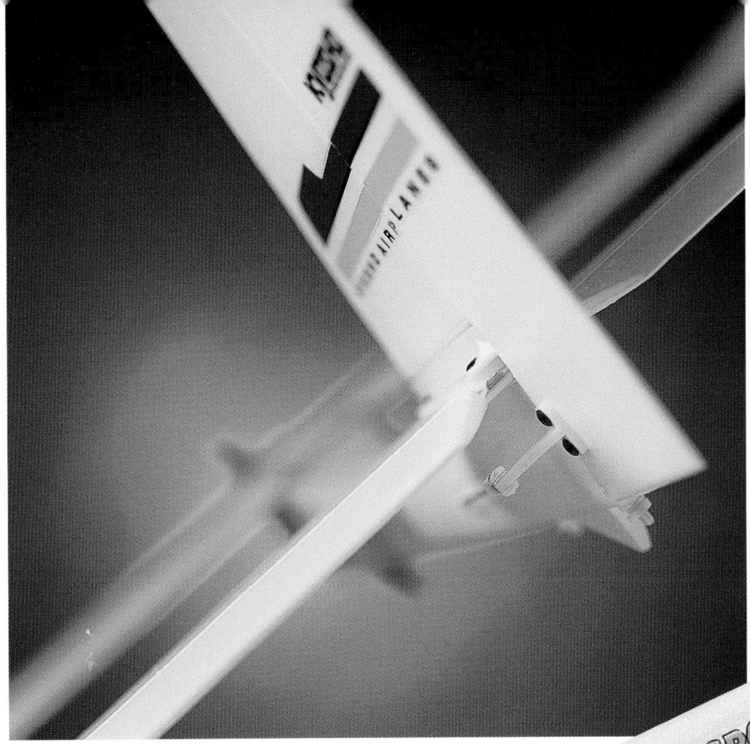

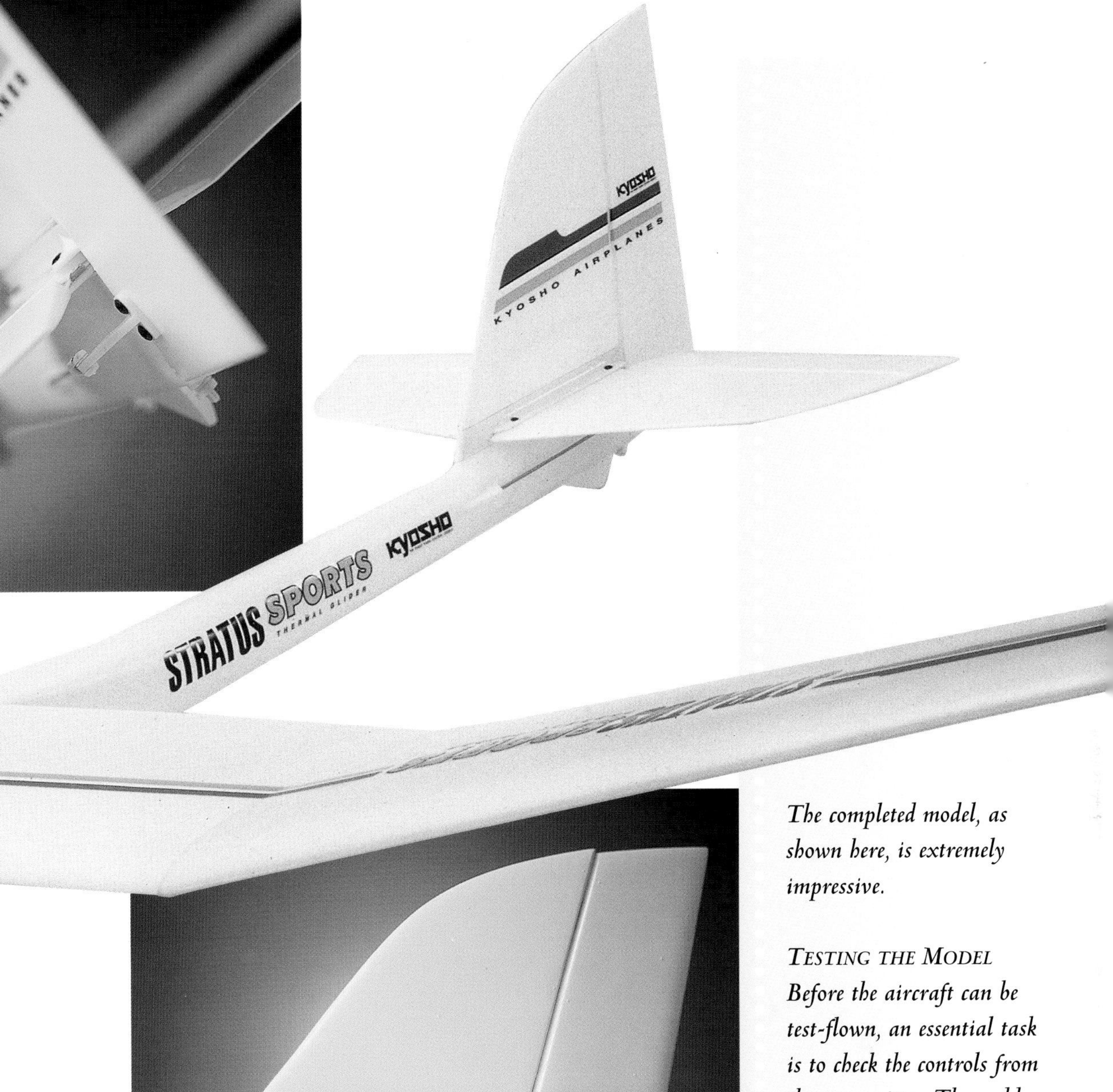

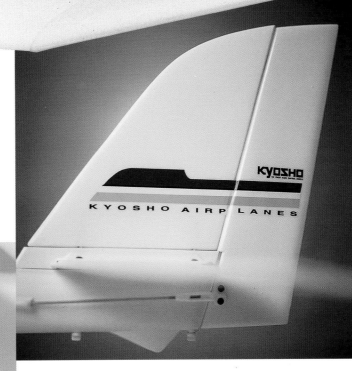

The completed model, as shown here, is extremely impressive.

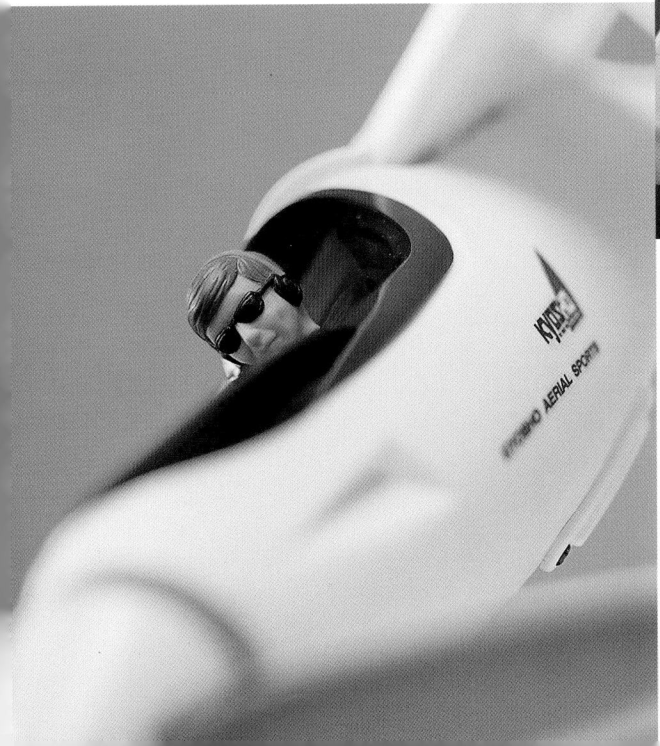

Testing the Model

Before the aircraft can be test-flown, an essential task is to check the controls from the transmitter. The rudder and elevator should move in response to the control stick, but the linkage may need to be shortened slightly by adjusting the threads on the ends of the pushrods, so that the rudder and elevator are straight when the control stick is centered.

See also

Control equipment, pp.58–59
Flying a glider, pp.60–61
Introduction to gliders, pp.56–57
Thermals, pp.58–59

Radio-controlled helicopters never fail to impress, because they can do everything a real helicopter can do, as well as spectacular aerobatics too risky for a real machine. However, of all radio-controlled models, helicopters are the most challenging to operate.

INTRODUCTION TO RADIO-CONTROLLED HELICOPTERS

A detailed knowledge of radio control and an understanding of the concept of helicopter flight are essential for mastering the skill of flying a model helicopter.

If you are a newcomer to radio-controlled helicopters, it is a good idea to start with the beginner's helicopter kits made by Kyosho. These helicopters are easy to build and easy to fly—one of them, the HyperFly, is featured on page 78. Although it is not strictly speaking a helicopter, since the tail rotor is replaced by a dished fin, the model enables a beginner to become familiar with the typical movements and control of a helicopter. Once you have made a HyperFly or one of its variants, you should have the confidence and experience to try one of the more complex helicopter kits on the market.

Helicopter kits are generally quite expensive, but the engineering involved is impressive. Models are available with both electric and IC motors. Some models are extremely large, and the usual safety and flying rules as noted on page 49 must all be observed .

ABOVE On a helicopter, the big rotor blades take the place of wings to provide the lift necessary for flight. As on a real helicopter, the rotor blades of a model need to be strong and well engineered.

RIGHT The realism and drama of rotary-wing flight is superbly demonstrated by this Robbe Futura Super Sport Trainer. This model is an ideal choice for a beginner.

A most attractive and realistic scale replica, the elegant Ecureuil 2 (Squirrel) from France has a fiberglass body that fits the Nexus 30 pod and boom.

CONTROLS

A model helicopter has a main rotor to give lift and movement, and a tail rotor to counteract the turning force of the main rotor. When the pitch, or speed, of the main rotor is changed, the balance with the tail rotor is upset, and the fuselage will start to move the other way. If the tail rotor malfunctions, the aircraft will autorotate and spin around out of control under the main rotor. The helicopter is moved up and down and to the side by continuously varying the pitch. Special transmitters are produced for model helicopters that have four channels controlling the necessary functions for flying a helicopter: left and right cyclic pitch; fore and aft cyclic pitch; yaw control; rotor speed and pitch. These functions are the same as for a real helicopter and allow you to back the machine left and right; pitch the machine backward and forward; turn left and right; and go up and down. Learning to control these movements simultaneously will enable you to fly the helicopter successfully.

To prevent them from yawing wildly as the rotor moves, helicopters are fitted with a gyroscopic stabilizer. This has a flywheel inside that detects any tendency to swing and automatically sends a signal to the servo controlling tail rotor pitch to change the pitch to compensate for the fuselage swing. This prevents yawing, although it does not necessarily hold the aircraft on its original course or axis. Some helicopters have a transmitter with a mixing switch that automatically alters the tail

RIGHT The transmission on the Nexus 30S, the latest in the Kyosho Concept 30 range.

ABOVE This close-up of the Kyosho Concept 60 pod shows the receiver covered with plastic sheet, and the gyro in the recess behind it; the plastic fuel tank is behind the gyro. The engine is not shown here.

ABOVE *The rotor head, which tilts the rotor blades and varies the pitch, is extremely intricate and well engineered.*

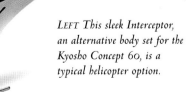

LEFT *This sleek Interceptor, an alternative body set for the Kyosho Concept 60, is a typical helicopter option.*

TRANSMITTER STICK MOVEMENTS
Use the transmitter switches as shown below to control the helicopter.

LEFT/RIGHT CYCLIC PITCH AXIS

FORE/AFT CYCLIC PITCH AXIS

YAW CONTROL

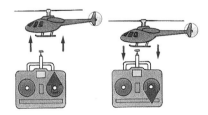

ROTOR SPEED AND PITCH

FLYING A HELICOPTER

Maximum concentration and good operating skills are required to fly a model helicopter successfully, because you will need to control at least four functions simultaneously. The best way to get flying experience and become fully proficient is to join a club where you can meet qualified instructors.

First practice takeoff, hovering, and landing. When you have mastered these key maneuvers, you can move on to forward flight and take your model around a circuit. When flying your helicopter, it is important to remember that every control movement can affect another function. For example, when you move into forward flight, the helicopter will start to lose height unless you open the throttle to balance the speed loss. This, in turn, however, may make the tail yaw, and you will need to be ready to correct this. However, with a little practice, anticipating the movements of your model will become much easier.

FLYING SAFETY

All the safety rules and operating tips for powered flight (see page 49) apply equally to helicopters.

RIGHT Ideal for experienced modelers, the Kyosho Concept 60 SRII is an extremely sophisticated model designed to meet competition flying standards.

BELOW It is important to carry out routine maintenance and safety checks on your model helicopter. Check that everything is tight and there are no signs of wear and tear. After each flight, check the linkages and the blades for signs of damage.

ABOVE Innovative in design, the Kyosho Nexus 30 has a muffler that can be adjusted according to choice of engine, and increased ground clearance for safer landings.

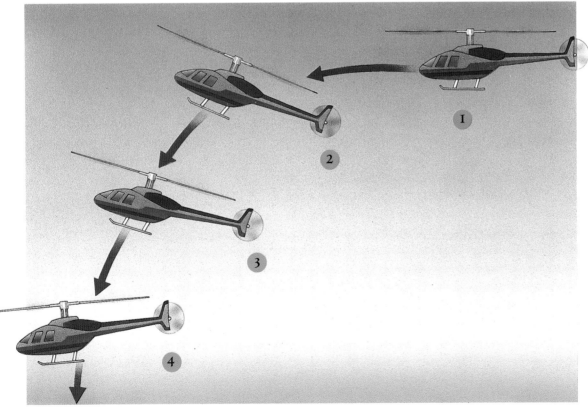

LEFT When landing your model helicopter, always follow this standard landing procedure.

1 Bring the helicopter toward the landing site.

2 Apply slight backward cyclic control and reduce power.

3 When the helicopter has slowed down, start the descent.

4 Just before touchdown, increase forward speed slightly to cushion the landing.

The Kyosho Hyperfly looks extremely realistic, yet is easy to build due to several innovative features. This model is also simple to fly as its special "dished" tail helps to prevent the model spinning out of control.

A CIVILIAN HELICOPTER

- **Easy screw assembly**
- **Matching control set**
- **Tail design prevents autorotation**
- **Military versions also available**

Radio-controlled helicopters are the most sophisticated models of all, demanding greater model engineering, patience, assembly skills, and flying skills than any other category of model. Many of the helicopter kits available are extremely complex and also more costly than other models. To avoid making an expensive mistake, be careful to choose a kit that is not beyond your experience and ability. An example of one of the simpler and less costly helicopter kits is the Hyperfly, manufactured by Kyosho, one of the leading producers of the more advanced helicopter models. The Hyperfly can be built in six to eight hours.

Simplifying features

The Hyperfly miniature helicopter looks realistic, yet is easy to build, and above all easy to fly. This was achieved by eliminating the most difficult elements of helicopter models, the gyro and the tail rotor. The Hyperfly looks as if it has a tail rotor, but it is actually a realistic cosmetic attachment. It is incorporated in a clever "dished" tail flying surface that keeps the model on course and prevents it autorotating, a common problem if the tail rotor fails. ▶

TOOLS YOU WILL NEED
Emery board
Craft knife
Small screwdriver

1 *Identify the control push-rods and attach them to the servo horns or rotor arms.*

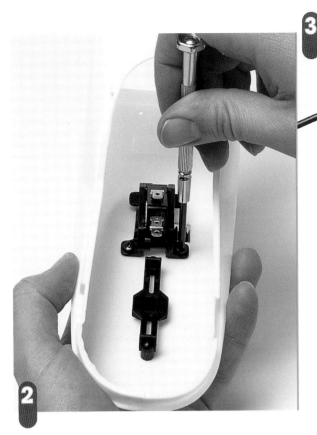

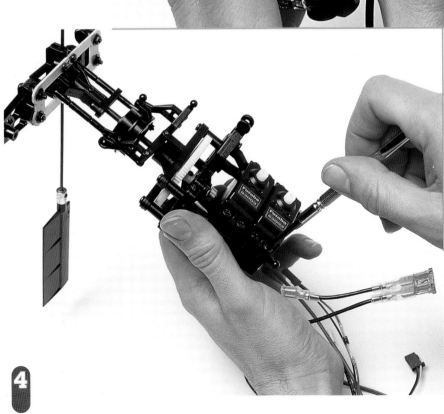

2 Find the mounting plate for the motor tower and then screw it firmly into place. This is easy, but do not forget to follow the instructions as you go along.

3 Assemble the electric motor into the drive and rotor housing and connect the shaft to the rotor head.

4 Complete the tower structure with the the motor, rotor head, shock absorbers, and power leads. This is largely a simple screw-assembly task

5 Now assemble the tail boom. This will complete the fuselage assembly, ready to take the motor and drive tower. Here, the collar for the support struts is being affixed in place.

Controls

Beginners can get the Hyperfly off the ground quickly and smoothly, as the model has a matching control set, the Futaba Attack. This control set has the servos and other equipment you need to get the model flying. The helicopter uses two-channel control and standard servos, although you could install lightweight servos. The standard servos are included in the special control set. Note, however, that you will need to buy the special Hyperfly battery pack. This special "square format" unit fits into the Hyperfly cockpit pod and is available from hobby stores that sell the Hyperfly kit.

Hyperfly variations

The Hyperfly is a colorful civilian helicopter. If you want a more aggressive-looking model, Kyosho produces two military helicopters based on the same basic pod and controls as the Hyperfly but with more complex body structures and extra detail. These military versions—the Hyperfly Apache, based on the US Army AH-64 attack helicopter, and the Hyperfly Manta, based on the futuristic McDonnell Douglas LHX Stealth helicopter—are made and flown in the same way as the Hyperfly, although the Manta has a more powerful motor.

6 *Assemble, install, and link up the receiver, switch, battery pack, and other control items. All the wires join as clip-fits, so there are no difficult connections.*

7 *Assemble the flybar paddles to the rotor head when the transmission/motor tower is built. Screw the sockets to the rotor head which will take the main rotors. These are clip-fits secured by screws.*

8 *Hold the transmission/motor tower up to the fuselage pod to check that it fits snugly on its mounting. Make sure that all the leads to be connected are stretched clear and are accessible and not tangled.*

Flybar paddle

Rotor head

Leads

Motor switch

9 *Secure the transmission/motor tower in place with screws through the side. No cementing or cutting is needed for this model. Everything either clips into place or is assembled with screws, making it a particularly easy construction task.*

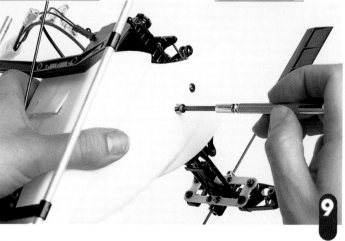

10 *Fix the tail boom, which also supports the radio aerial, to the socket at the rear of the transmission/motor tower. Note the wire aerial set out behind the fuselage pod. It will later be held by the tail boom supports.*

11 *Next add the tail boom supports. They are screwed to lugs in the top of the transmission/motor tower and secured on the tail boom where they also hold the aerial in place when it is fixed, and the rear end of the boom.*

12 *The foam core rotors are very strong. When all the tail boom rigging is complete, attach them to the special lugs on the rotor head. At this stage the battery can also be inserted and wired up. Then turn on the transmitter and test the controls.*

13 *Trim and clean up the transparent cockpit canopy and then test it for fit. It will be attached later with double-sided tape to allow easy access to the battery which fits inside the cockpit. You will also need to affix the framing decals to the cockpit canopy.*

14 Using small scissors, cut the self-adhesive decals into individual pieces and fix them in the appropriate position. Rub them over with a damp, slightly soapy cloth to eliminate any air bubbles.

15 Check that all the fittings are secure and that the skid undercarriage is correctly assembled and balanced. Also make sure that the important auto cut-out switch for the motor moves correctly.

NOTE

Not shown here is some preliminary work needed before assembly can begin. This entails cutting away a "solid" plastic molding from the cockpit area and rear of the fuselage pod. This requires great care, because the plastic is thick and hard and the knife may slip. Trim the cockpit transparency to size and test-fit it as work proceeds to check you have cut away sufficient plastic.

The completed model is now ready for flight. Note the cosmetic but effective dummy "dished" tail rotor that acts as a stabilizer in flight. The radio aerial can be seen running below the tail boom.

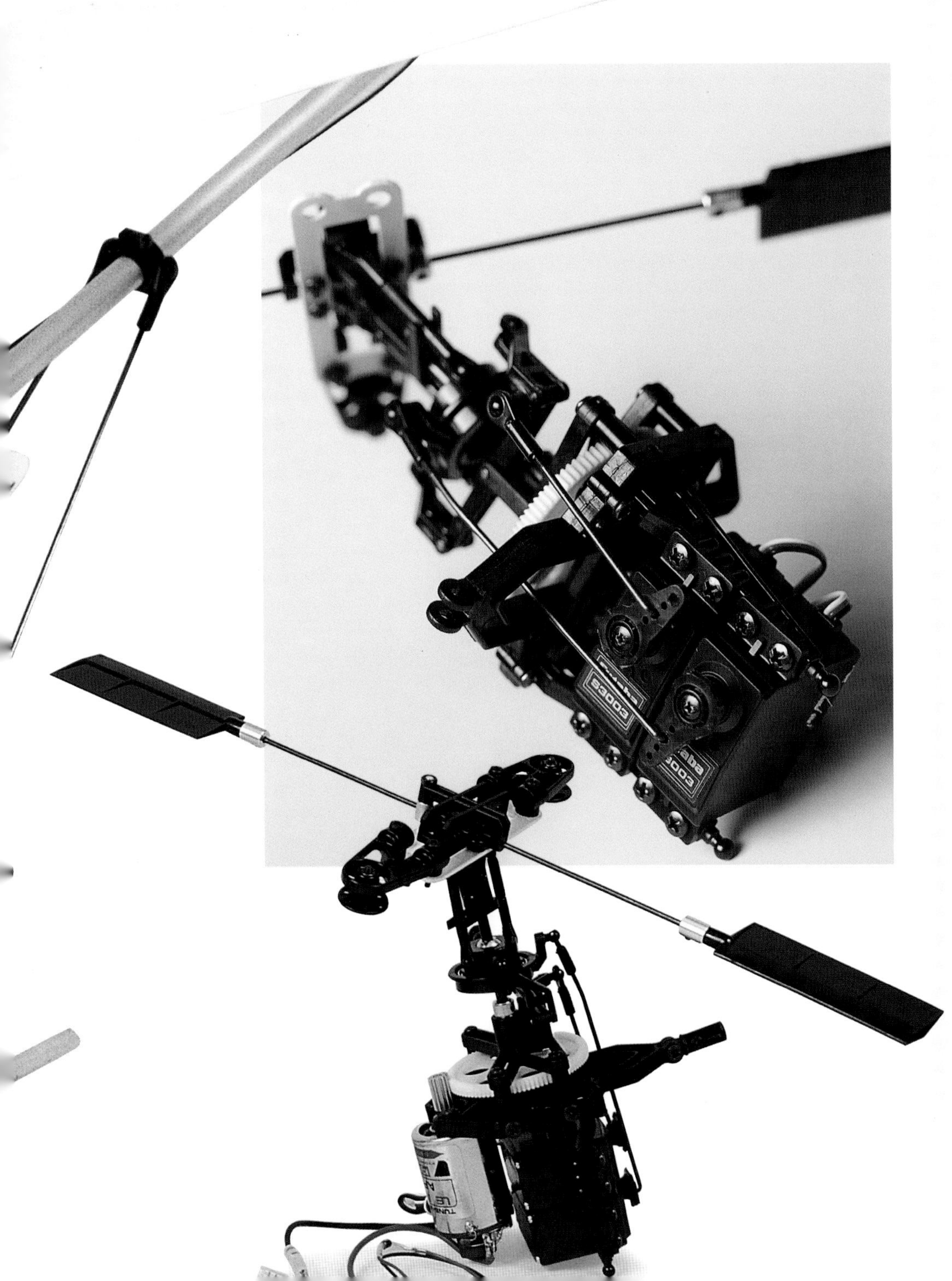

FLYING THE MODEL

The Hyperfly has particular flying characteristics that are slightly different from those of the more advanced helicopters with gyro stabilization. The Hyperfly must be hand-launched and only flown in light winds, not gusty conditions. Clip in the battery, switch on the transmitter, and start the motor by pulling down on the motor switch. When the motor is running, lower the motor switch to its fullest extent. Then launch the Hyperfly forward and upward, keeping the nose down at about 15° to the horizontal. Let it fly out of your hand—do not throw it forward. It is best to work with a friend, at least for early flights, with you handling the transmitter and your friend doing the launching. Later you may become skilled enough to launch the Hyperfly with one hand while handling the transmitter slung from your neck with the other hand. It is best to fly the Hyperfly into the wind and make turn maneuvers nose to wind, where possible. A strong tail wind may cause the Hyperfly to autorotate. If this happens, pull down on the elevator control to restore straight flight. You can bring the Hyperfly into land on the ground or any other flat surface. When the motor switch below the undercarriage skids touches the ground or any other hard surface, the whisker is pushed up, cutting out the engine instantly as the helicopter touches down.

SEE ALSO
Controls for helicopters, pp.74–75
Introduction to helicopters, pp.72–73
Landing a helicopter, pp.76–77

Of all radio-controlled models, airplanes, gliders, and helicopters require the greatest skills, not only to build but to operate and maintain. However, there is a wide range of model aircraft kits, ranging from simple gliders to complex scratch-built aircraft. This section shows you some of the possibilities available at levels ranging from beginner's models to complex kits for the advanced modeler.

GALLERY

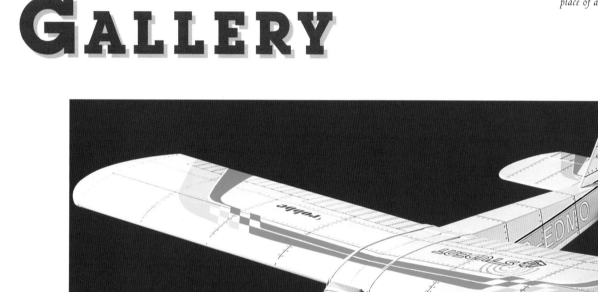

LEFT *The semi-scale F-15 Eagle, by Great Planes, looks extremely realistic, but it has a simplified fixed undercarriage and a front-mounted IC engine and a propeller in place of a jet engine.*

ABOVE *The Robbe Student V is a particularly compact Trainer airplane that is easy to assemble. All the main components are pre-formed and colored. The model comes with a small IC engine and fuel tank. A similar model, the Student E, has an electric motor.*

RIGHT *This highly detailed scale flying model is a replica of the US Navy Douglas Skyraider used in the Korean and Vietnam wars.*

ABOVE *These Robbe models of the famous World War II fighters, the Supermarine Spitfire and the Messerschmitt 109F, are made of styrofoam. The Spitfire wingspan is 44in (110cm) and the Messerschmitt 109F wingspan is 41in (102cm). The models have similar assembly and controls to the Airdancer featured on page 50. They are designed for "dog fighting" events, which have recently become popular in the US. In these fighting competitions, the models are hand-launched with a streamer that can be cut by an "attacking" aircraft. The models have a special hand recess below the fuselage for launching.*

RIGHT This spectacular Kyosho Etude Trainer model is fitted with floats, enabling it to land on and take off from water. Note the fitting of a "snake" flexible control linkage from the airplane rudder to the float rudder, which allows it to be steered on water.

ABOVE The Spot is a hand-launched glider specially designed by Robbe for newcomers to radio-controlled gliding. It incorporates many recent aerodynamic features and comes with ready-made epoxy wings and fuselage. The control equipment is included in the kit, making assembly and flight quick and easy.

RIGHT The Robbe Pioneer is a traditional wooden glider with a 74in (185cm) wingspan. The model has a box section fuselage for strength and simplicity, and is covered with plastic film. Designed for two-channel radio control, it is a high-performance model. An electric version, the Pioneer ARC, is also available.

BELOW *The Robbe Futura is one of the most sophisticated helicopters available. Introduced in the mid-1990s, it has won many European championship trophies for model helicopter flying. Modeled on a helicopter used by the German national helicopter team, it has a 10cc IC engine, rotor diameter of 60in (149cm) and a length of 62in (154cm). This model is fitted with the optional larger 72in (180cm) rotor, a more powerful Super Tigre engine, and extra tuning parts.*

ROAD VEHICLES

Radio-controlled road vehicles are some of the simplest models to build and operate. The high performance and quality engineering of the cars featured here will appeal to beginners and experienced modelers alike.

Model cars are the most popular and most accessible of all radio-controlled kits. As a result, the market is very active, with numerous new models coming out every year.

INTRODUCTION TO RADIO-CONTROLLED CARS

The popularity of radio-controlled cars is easy to explain. The leading brands of car are familiar to most people, and racing cars have a high profile in the media. As a result, working miniature reproductions of these cars that are superbly engineered are guaranteed to receive an enthusiastic reception. In addition, frequent technical developments, such as improved suspension and drive systems, ensure that the leading brands of radio-controlled car are reliable and efficient.

The fact that many radio-controlled car kits are easily assembled (the parts need only be bolted together) does not detract from the satisfaction of building them. The simplicity is balanced out by the sheer pleasure of assembling a well-

engineered model that looks good and works well when finished. In addition, these models are easy to operate. Unlike radio-controlled aircraft or boats, which need plenty of space to operate and suitable facilities, model cars are compact enough to operate in your backyard or driveway. Radio-controlled cars also require less setting up than aircraft and boats—all you need to do is recharge the battery and do routine checks and you are ready to operate. If you are a complete beginner to radio-controlled models, a simple model car kit makes a fine introduction to the hobby. Car kits and all the control equipment you need are available at all model stores and some department stores.

ABOVE The Tamiya Mini-Cooper made up from the kit, with added control gear on the chassis. The drive is to the front axle, as in the real car. The model is decorated with Monte Carlo Rally trim.

LEFT *This Kyosho Corvette Mantis is a good example of the very fine finish possible with a painted ABS body.*

BELOW *A typical modern car kit, such as this Tamiya 1:10-scale Mini Cooper kit, has many parts. Note that the ni-cad battery and receiver unit shown here are not always included with the kit.*

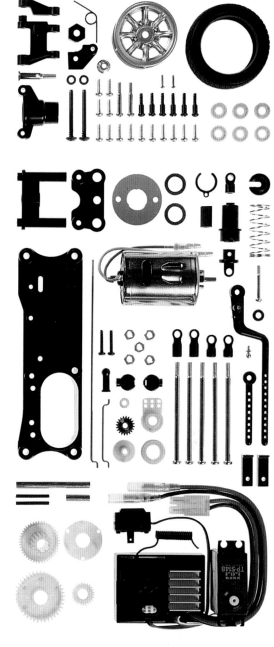

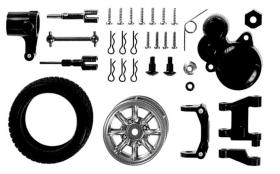

SCALE

Model cars come in a wide range of sizes. All models are designed to scale and in some cases the most popular types are available in several different scales. Some models, such as those used in radio-controlled slot car racing, are as small as 1:42, while others are 1:4 scale—an enormous quarter full-size. The most common scales, however, are 1:10 or 1:12 (1:14 or 1:16 for trucks). Models in these scales are a good size and show plenty of detail without taking up too much room. The larger the scale, the more space the model needs and the more difficult it is to carry around. Although the scales are decided mainly at the whim of the kit maker, in some cases they may be influenced by the need to use common components or by competition classes. If there are competitions for a model of a certain scale, you can usually find kits in that particular scale.

ABOVE This Kyosho model is a 1:10-scale replica of the classic Ferrari 330 P4.

BELOW With its sponsor stickers for rallies, this Kyosho Porsche 911 is a fine example of the spectacular finish possible on GT cars.

MATERIALS AND COMPONENTS

Some simple RTR model cars have hard plastic (polystyrene) bodies but most kit-built cars are molded in ABS (Lexan). This material is flexible yet immensely tough. Most ABS car shells are molded in one piece and need to be painted, since they are transparent. Polycarbonate paint is produced specially for ABS and this is applied on the inside. For a perfect hard-gloss finish you may need two or more coats of paint. Protective coatings, which prevent paint discoloration, or deterioration due to fuel spillage, can be applied around the engine area of IC-engine models. The chassis and suspension construction is usually made of hard, tough plastic, but fiberglass reinforced plastic and carbon-graphite are also used. Some of the latest high-performance cars have an aluminum chassis, which is light yet strong. An interesting idea is to buy a standard high-performance chassis that fits most GT types (wheelbase adjustable) and use it with a variety of bodies.

ENGINES

Model cars are usually powered by an electric motor. These have no noise or fumes, making them environmentally friendly. Some models may have more than one motor, but this is rare. Model cars may be two-wheel drive (2WD) or four-wheel drive (4WD). The former are most often driven via a gearbox on the rear axle, but where appropriate (or true to type) may be driven on the front wheels. The latter are driven on each axle, front and rear. Power may be transmitted by a shaft front and rear, as with a real car, or by a belt driven from the engine. A few 4WD model cars may have a motor driving each axle.

The alternative to an electric engine is an internal combustion (IC) motor, sometimes known as a "nitro" or "glow" engine. This engine is a masterly precision reduction of a real engine, but single cylinder only. However, it has the features of a real engine, such as a glow plug, needle valve, piston, crank shaft, clutch, carburetor, starter, and muffler (silencer). IC-engine cars are usually to 1:8 scale or larger. Most IC engines are air-cooled for simplicity so you will find the characteristic cooling gills around the cylinder. (Water-cooling is mostly found on powerboat models, where the water comes from the pond itself.) The operation is 2-stroke—virtually the same as a full-size engine. Most engines are started by a pull-start that pulls against a recoil spring to start the crankshaft moving and cause the cylinder to combust. Some models have an electric starter, however.

The clutch comes in when the revolutions reach operating speed, determined by operating the throttle via the servo. A brake unit is synchronized to the throttle servo, and this is activated when the engine power is reduced by the same throttle servo. This gives a realistic braking effect when you slow for a corner, for example. IC engines use a special fuel, available at hobby stores, containing methanol and nitromethane; no other fuel should be used. A typical model car can carry enough fuel in its tank for 8–10 minutes running. Although the noise and power of IC engines is attractive to modelers, they are not environmentally friendly, in spite of modern developments in cylinder heads and mufflers. If you choose an IC-engine car, check that there are no operating restrictions in your area. Alternatively, run the car on designated circuits.

ABOVE The IC engine is visible in this Kyosho ARTR (almost-ready-to run) Volvo 850.

BELOW Pull-starting an IC engine.

ABOVE The Pure 10 GP chassis is a high-performance IC-engine unit. Made of aluminum, it is extremely light and can be used with different bodies. A ventilated disk brake is included in the design.

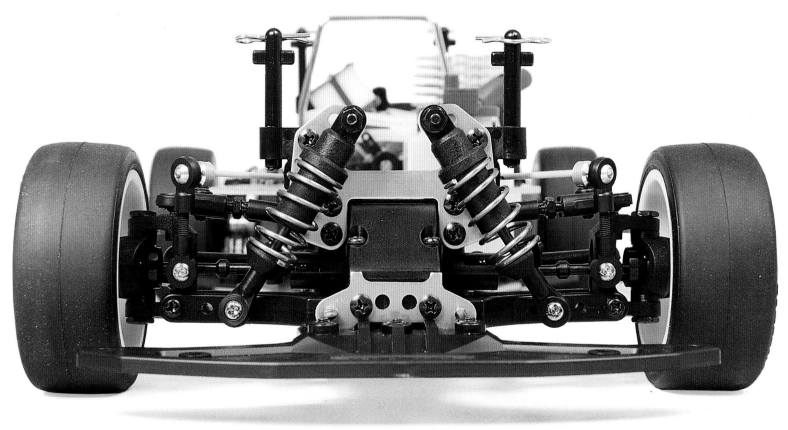

ABOVE *The strengthened suspension and slick tires on this Formula 1 GP Spider by Kyosho are typical examples of "hot-up" option parts.*

BELOW *These hot-up racing-wheel hubs provide a lightweight alternative for a model car.*

TUNING

Tuning is the art of improving your car's performance with enhanced or special "hot-up" or "hop-up" components. A variety of tuning methods are possible. For example, you could replace plastic or metal bushings with ball bearings; use lighter parts made in aluminum; use extra shock absorbers and coil spring dampers; use spring-loaded high-torque (faster-response) servos in place of the standard units; reinforce the chassis; replace the standard motor with a competition (higher-revving) motor; use "racing" battery packs; use a lighter flywheel on an IC-engine car; and use "competition" fuel with a higher nitro-methane content in IC-engine cars. All the leading model-car makers, such as Tamiya and Kyosho, supply tuning parts for better and faster performance, such as different wheels and tires, better suspension, tighter springs, sport-tuned engines, alternative gears, lighting sets, ball bearings, and adjustable body mounts. Some kits have built-in tuning options. For example, the simple Fighter Buggy featured on page 102 has different suspension settings for smooth surfaces and rough terrain.

RADIO CONTROL FOR CARS

Most radio-controlled cars use two-channel radio control. The transmitter is usually the straightforward, standard "box" type. The left-hand stick moves up and down to control the speed, and the right-hand stick moves left and right to control the steering. An alternative favored by many experienced modelers is the hand-held "pistol grip" wheel-type transmitter. The steering is controlled by a wheel on the side of the unit and the speed is controlled by pressure on the "trigger." This type appeals in particular to those who like to steer with a wheel.

Like all transmitters used in radio control, model-car transmitters have a crystal match—one in the transmitter and a matching one in the receiver, color- and number-coded. However, the crystals in model cars are particularly important. When several cars are running together, whether racing formally or just running informally, each car must have its own unique frequency determined by the crystal set in order to maintain control discipline. If you are running your car with a group, you may have to change crystals so that no two modelers are operating on the same frequency. Even if all the crystals have been checked, always switch your transmitter off, if you are not actually running your car. A mismatch, caused, for example, by a spare car in a bag that you do not know about, can cause chaos and control loss.

The remaining control elements—the receiver, servos, motor, and battery—are found in the car. The aerial is also important, since without it the signals cannot be received. Before operating, always check that the aerial is corrrectly in position and connected, since it is easy to push the aerial out of position, or buckle it, when re-fitting the body after checking the motor.

Once received via the aerial, the signal is distributed by the receiver via connecting leads to the steering servo and the motor servo. The steering servo transmits the physical movement of the steering wheels, and the motor servo controls forward/stop/reverse on the motor. With an IC engine, the latter operates the motor throttle. Some radio controls have a mechanical speed control unit whereby the motor servo controls a resistor unit that regulates current flow to the motor in a number of steps—fast, medium, and slow speeds. In effect, this is a model equivalent of a gearbox in terms of speed, though no gear shifting is done, of course. However, nowadays, all but the cheapest models have an electronic speed control unit. These units have transistors which use amplification to control the current supply to the motor to give a similar fast, medium, and slow effect, but with a smoother transition up and down the speed range. The unit may be built into the receiver or it may be a separate unit.

Radio-controlled cars generally use a rechargable ni-cad battery pack, a cell-type unit

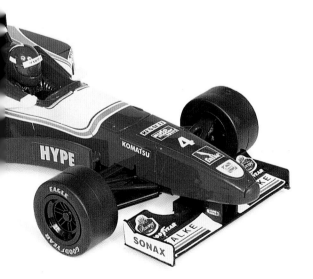

ABOVE **The difference between a car with mechanical speed control (above) and one with electronic speed control (right).**

Battery

Motor

Resistor

Steerable front axles

Steering servo

Receiver

Speed control servo

Receiver and electronic speed control

Steering servo

Battery

Motor

LEFT **This Renault F1 GP car, 1:10 scale by Kyosho, is shown with its box-type transmitter.**

with leads built in. Several different sizes and outputs are used, depending on the model. An essential extra accessory is a battery charger. This connects to the main supply and is used to charge up an exhausted ni-cad battery pack for the next session. A "quick charger" is also available. All the main manufacturers of car kits produce battery chargers and battery packs for their models. In practice most makes work with others, but for guaranteed compatibility it is a good idea to use the accessory items from the same range as your model kit. Always remove the battery pack whenever the car is not being used and recharge flat battery packs after each operating session. To keep your car, or cars, running you may need several battery packs to hand.

Technical innovations for radio-controlled cars are many and varied. For example, the latest Nikko 1:8 scale Mercedes-Benz ITC GT car has a starter button on the transmitter which allows the IC engine to be started remotely from the transmitter instead of by a pull-start on the car itself. Other vehicles use a third channel or special function switches to perform extra functions, such as shifting gear, opening the trunk, operating the head and tail lights, and adjusting the suspension.

MAINTENANCE

Model cars can get dirty, and suffer general wear and tear. However, careful maintenance can help to prevent your car from malfunctioning. After every session it is important to examine the chassis and body separately. Check the body for paint damage or cracks, and repair if necessary. Clean the chassis, check that all the nuts and bolts and fittings are tight and intact, and lubricate everything that moves and needs lubrication. Replacing any bent or damaged parts is easy, since all parts are sold as spares. (Remember to write down the numbers of the parts before you go to the store to avoid mistakes.) Check the steering linkage and suspension regularly to prevent future damage. If you do make any adjustments, replace the battery and check the control functions again. Make sure that the steering lines up, so that the wheels are straight when the stick is centered. If the steering is not functioning properly, the servo linkage or the servo spindle itself may be worn, and you may need to replace the servo and/or the linkage.

TOOLS

Apart from general modeling tools, you will need a box wrench, a screwdriver set with both straight- and cross-head ends, Allen keys in several sizes, and long-nosed small pliers. Liquid thread lock, to ensure that key screws do not come loose, is also useful. Some tools, such as box wrenches and Allen keys, may be supplied with the kit. It is always a good idea to keep these in your tool set after the model is finished, since they will come in handy again.

RIGHT Test an IC-engine car by firing up the cylinder with a Glo-Starter. This convenient start-up device has a rechargeable ni-cad battery in the handle

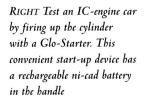

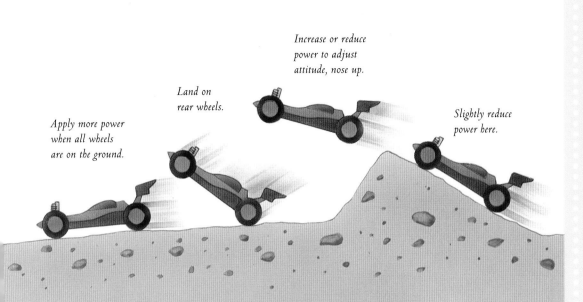

Increase or reduce power to adjust attitude, nose up.

Land on rear wheels.

Apply more power when all wheels are on the ground.

Slightly reduce power here.

DRIVING

A degree of skill not unlike that needed for driving real cars is necessary for driving a radio-controlled car. Typical maneuvers, such as cornering, turning, Y-turns (three-point turns), and overtaking, all need to be undertaken with the same care as if you were driving a real car. A key tip for safe and accurate model-car driving is to watch the path immediately ahead of the car, and not the car itself. By doing this you will be able to see corners, turns, and obstacles before the car reaches them. You can test your driving skills by setting up course markers at your running site. Make markers from everyday items, such as upturned plant pots, or buy bright orange course markers from Tamiya. You can practice backing and turning and weaving through the row of markers, and also hold competitions with friends.

ABOVE You will need superior operating skills to run a model buggy successfully on an off-road course. If you want to extend your skills further, try building a cross-country circuit for off-road cars and buggies in your yard. You can arrange a number of obstacles around a rockery, for example.

DOS AND DON'TS OF MODEL DRIVING

● Keep off public roads and away from any real cars.
● Avoid bystanders, passers-by, and pets and other animals.
● Check noise restrictions if you are running an IC-engine car.
● Only operate indoors if there is sufficient space to maneuver the car.
● Do not run full speed into the curb, furniture, structures, or people. Anything that brings the model to a sudden stop may damage or ruin the motor, controls, or structure of the vehicle.
● Check your frequency (crystals) with any other car modelers operating nearby.
● Always switch off the transmitter when your car is not running.
● Check your car after running, and clean and repair any damage.
● Remove the battery pack after an operating session and recharge.
● Never touch the motor after running, because it may be hot enough to burn the skin. Allow it to cool down first.
● If you are operating the car in your yard or driveway, make sure other family members know, so that they do not walk into the path of your car.
● When jumping a buggy or off-road vehicle off a ramp onto rough terrain, keep the car straight, so that it lands squarely, and keep the power up sufficiently to raise the nose, so the car lands on its rear wheels.
● Only apply power when all the wheels are in contact with the ground.
● Avoid soft loose surfaces, such as sand, where possible. If you do run onto soft ground, drive gently to stop the wheels spinning deep into the surface and stalling the car.
● Keep out of puddles, water, or rain, where possible. If the model does get wet, dry both the inside (controls) and outside (bodywork) at once.
● Look all around and ahead before driving off.
● Be aware and concentrate at all times to make sure none of the above problems is encountered.

RACING

Racing your model car is an even greater test of your driving skills. The best way to start radio-controlled car racing is to join a club specifically for radio-controlled car enthusiasts. There are an increasing number of such clubs—some even have their own racetracks, complete with raised stands for drivers and spectator seating and other facilities. To find a club near you, check out current model-car magazines. In addition, national organizations, such as the North American Racing Radio Control Association (NARRCA) and the British Radio Car Association (BRCA), sponsor championships and draw up rules for classes and competitions. They are also a useful source of information about clubs and the hobby in general. Their addresses can be found in model-car magazines. There is also an international body that organizes some world championships. In addition, the leading radio-controlled model car firms, such as Tamiya and Kyosho, organize or sponsor both national and international championships for models made from their kits. You may also find that your local hobby store or hobby distributor occasionally organizes meetings and championships.

Some of the best racetracks are miniature versions of real race circuits and are very impressive. Race meets are strictly organized as there may be many events for several different classes of car. Although there are stewards to ensure the races run smoothly, it is vital that you know your start time so you do not hold up the events. You will also find pits for servicing and fueling, a race control center with briefings and (usually) refreshments, and sometimes even a trade stand and sales booths.

You can also build your own race circuit if you have access to sufficient private land. Alternatively, you may be able to get permission to use a school yard during vacations. The *Radio Control Guide Book*, produced and updated regularly by Tamiya, gives detailed information on racing, and building your own race circuit. Tamiya have extensive experience of holding competitions, and details of several race circuits are given. A race circuit recommended by Tamiya is shown here to give you an idea of how you might design and organize your own circuit.

ABOVE *A competitor gets ready to race his model at the start line-up at a radio-controlled car-racing championship organized by a leading kit manufacturer.*

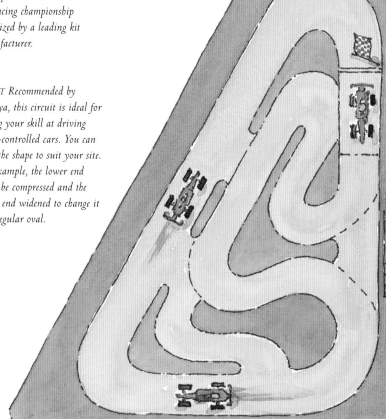

RIGHT *Recommended by Tamiya, this circuit is ideal for testing your skill at driving radio-controlled cars. You can vary the shape to suit your site. For example, the lower end could be compressed and the upper end widened to change it to a regular oval.*

ABOVE These two Tamiya cars in action on a race circuit demonstrate the thrill of 1:10 scale radio-controlled model racing.

RACING CLASSES

Due to the popularity of racing radio-controlled cars, there are now a number of racing classes which cover most types of radio-controlled vehicles. There are also classes specifically for IC-engine cars. In addition to the classes below, there are a few classes unique to a maker or country, such as the Mini Stock 1:12 scale class for indoor racing, introduced by the British company of Margrave.

Radio control racing

Developed in recent years by Nikko, "radio control racing" is an updated version of traditional slot-racing. The cars still pick up power from a slot in the track but speed and maneuvering is controlled by a transmitter instead of a hand throttle wired to the track. With the new system, lane changing and overtaking are possible. The models used for this type of racing are 1:42 scale—the same size as small die-cast model cars.

RIGHT This bathtub-type chassis is used in many 1:10 scale Tamiya Touring class cars. Note how the battery pack fits neatly in a recess below the control gear tray.

Formula 1

This class is for models of Formula 1 racing cars, built to 1:10 scale.

Touring

Also known as Tamiya Touring, this class was introduced by Tamiya, manufacturers of a popular range of GT cars of the type seen in rallies and GT-class racing all over the world. The scale is 1:10, and models may be in 4WD or 2WD.

Pick-ups and big wheels

This class is for models of pick-up trucks, including the "big wheel" types. These models perform extremely well off road and are a lot of fun to operate in competitions. The scale is 1:8, 1:9, or 1:10, and models may be 4WD or 2WD.

ABOVE Ultrex by Kyosho is typical of 1:10 scale buggies for off-road racing.

Buggies

This class is for the "dune runner" type of beach cars that were originally popularized in the United States but are now known all over the world. Racing these off-road cars over rough terrain is exciting. This is probably the most popular class, because the models are tough and there is not much bodywork to get damaged. The scale is 1:10 and models may be in 4WD or 2WD.

LEFT This Kyosho "big wheel" pick-up truck performs spectacularly in stadium track racing, with its beefed-up high suspension.

RTR (ready-to-run) and ARTR (almost-ready-to-run)

Also known as RR or ARR, these models look like and even run like kit-built models. The more sophisticated RR models are ideal for those who are interested in radio-controlled cars but who might not have the time or skill to make a car from a kit. ARTR models come part-assembled and are useful for those with limited time or skill. The motor and drive are usually already installed in a made-up chassis, so all you need to do is add the body, details, wheels, and so on.

Stock and Modified

These are categories that you will encounter if you get involved in races organized by clubs. Stock cars are those built using the standard motor and kit parts, although you may improve your model by polishing and lubricating, etc. Modified cars are those freely altered by using "hot up" parts, such as more powerful motors.

STADIUM TRUCKS

Trucks in this exciting American racing class are modified pick-ups with beefed-up suspension and big wheels, in 1:10 scale.

MOTOR CYCLES

An interesting development in recent years is the radio-controlled motor cycle. These were pioneered by Kyosho with their "Hang On" series. You may be surprised to see that they run just like real bikes, with no outriggers or other devices to keep them upright. They have chain drive to the rear axle and realistic steering, with the driver leaning over on bends. They use two-channel control but have to use a mini-servo and a combined unit carrying the receiver, speed controller, and steering servo all in one, for space reasons. Option and tuning parts are also available. The scale is 1:8.

RIGHT Radio-controlled motor cycles, such as this Suzuki RGV by Kyosho, are exciting new products from model manufacturers.

SAFETY RULES FOR CAR RACING

Most racing rules, including the use of a checkered flag to indicate the winner, are based on those used on full-size circuits. Key points include:
● Spectator areas must be protected by light fencing at least 2ft (60cm) high. On a temporary circuit, use a row of chairs or move spectators well back from the edge.
● Model cars can get very hot. Always allow the motor to cool before handling.
● Switch off your transmitter when your car is not in use.
● Only stewards are allowed to enter the circuit area.
● Fill out any necessary registration or entry forms ahead of time to avoid holding up the schedule.
● Check your starting time and that you know your operating position.
● Check with the race stewards that you have the correct crystal for your race.
● Find the pit and refueling area and check the admission rules.

An ideal model for complete beginners, the Fighter Buggy RX, manufactured by Tamiya, has a simple control hook-up. In spite of its simplicity, it performs well and serves as a good introduction to radio-controlled cars.

OFF-ROAD 2WD BUGGY

- *Ideal for beginners*
- *Easy to assemble*
- *Lively off-road performance*

If you have never tried to build a radio-controlled model from a kit before, it's a good idea to start with the Fighter Buggy RX. This model is one of the simplest off-road, 1:10 scale class, 2WD (two-wheel drive) vehicles. The 2WD format avoids the assembly complications.encountered with 4WD (four-wheel drive) models. Note that even for this beginner's model there are a number of "add-on" or "hot-up" options available.

Getting started

Like all kits of this type, the Fighter Buggy comes in a large box with all components clearly labeled and small parts carefully packaged. Resist all temptation to take the parts out of the box before you read the instruction sheet and check that all the parts are included. Always keep the small parts in the kit box to avoid losing any important pieces.

YOU WILL NEED

Small screwdriver set including
  Phillips crosshead
Allen key
Small spanner
Craft knife
Small file or emery board
Plastic and contact adhesives
Small scissors
Side cutter or small pliers
Box wrench
(Some of these tools will be in the kit, e.g. Allen key)

Assembly

If you are new to kit construction, even a fairly simple model like this may look complex. Before you begin, make sure you understand the basic assembly steps listed in the comprehensive instruction book supplied with the kit.

1 *Check all the plastic-molded parts for any slight defects. These may include visible "parting lines," "flash" (excess plastic on edges), or other mold marks. Here a ragged edge on the spoiler is cleaned up and smoothed down using a side cutter and file.*

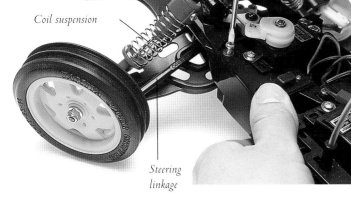

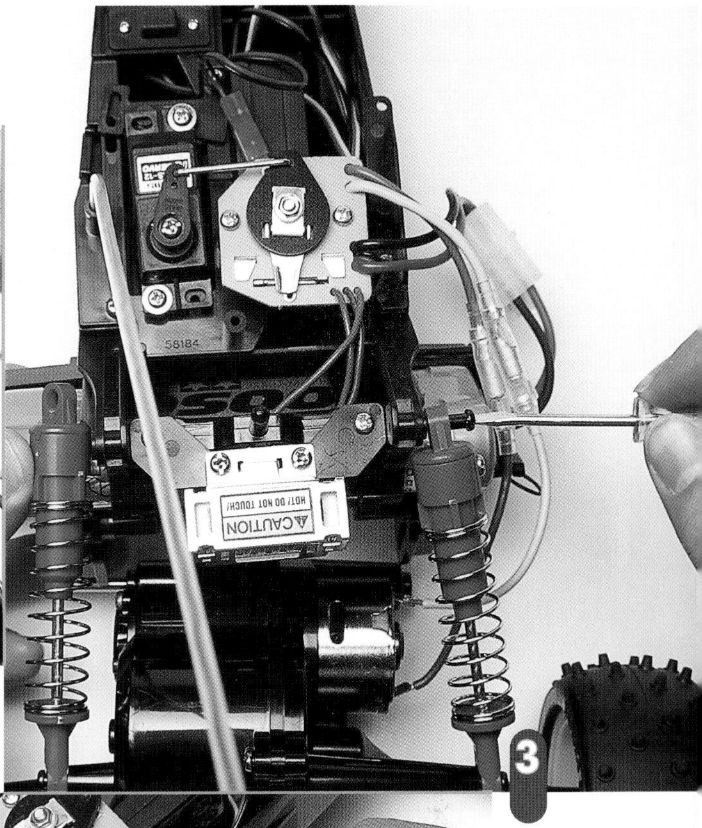

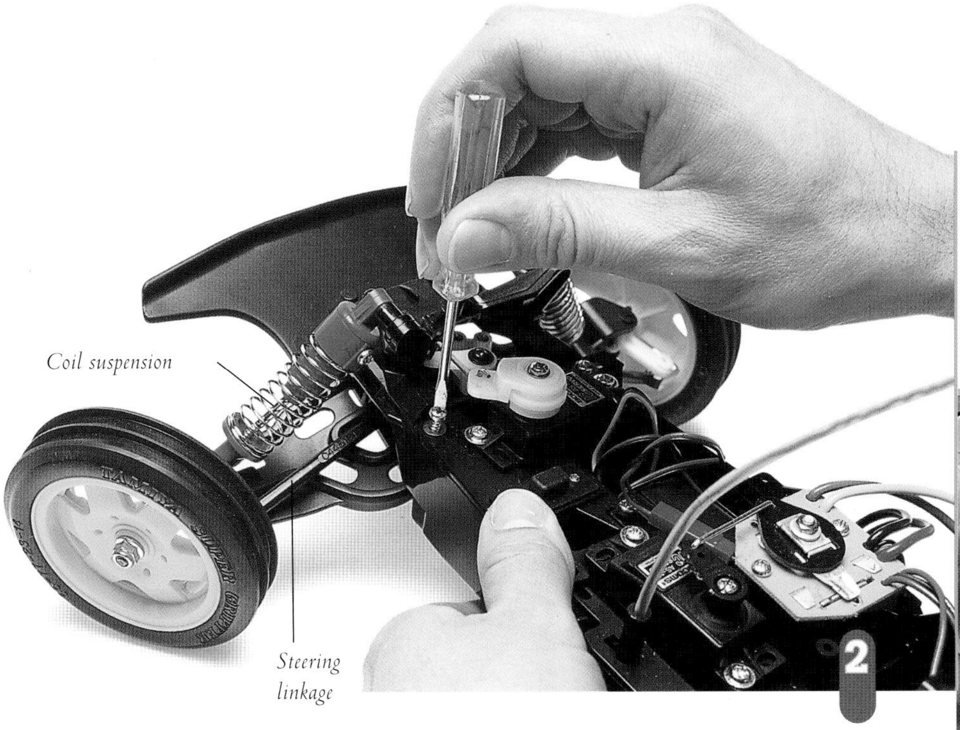

Coil suspension

Steering
linkage

2 Though the finished chassis looks quite complicated, it is easy to assemble in stages. After completing the front suspension, screw the complete subassembly into place on the front of the chassis pan.

3 Fix the rear suspension unit and gearbox to the rear of the chassis pan in the same way. Clip the cable lead from the battery housing in the chassis to the motor leads for the electric motor, which is positioned immediately in front of the gearbox. On top of the chassis is the electronic speed control box, which is positioned above the receiver unit.

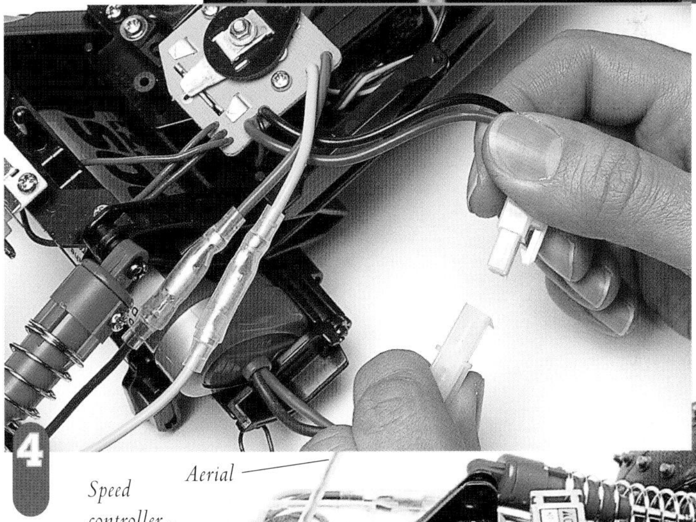

4 The task of joining the electronic speed controller to the motor is made easy by using standard insulated plastic-covered connector plugs. Here, the ni-cad battery is being connected.

5 The assembled Fighter Buggy RX chassis is now ready for testing. Note that the aerial is attached. When the body is in place, the aerial is taken out through one of the side windows.

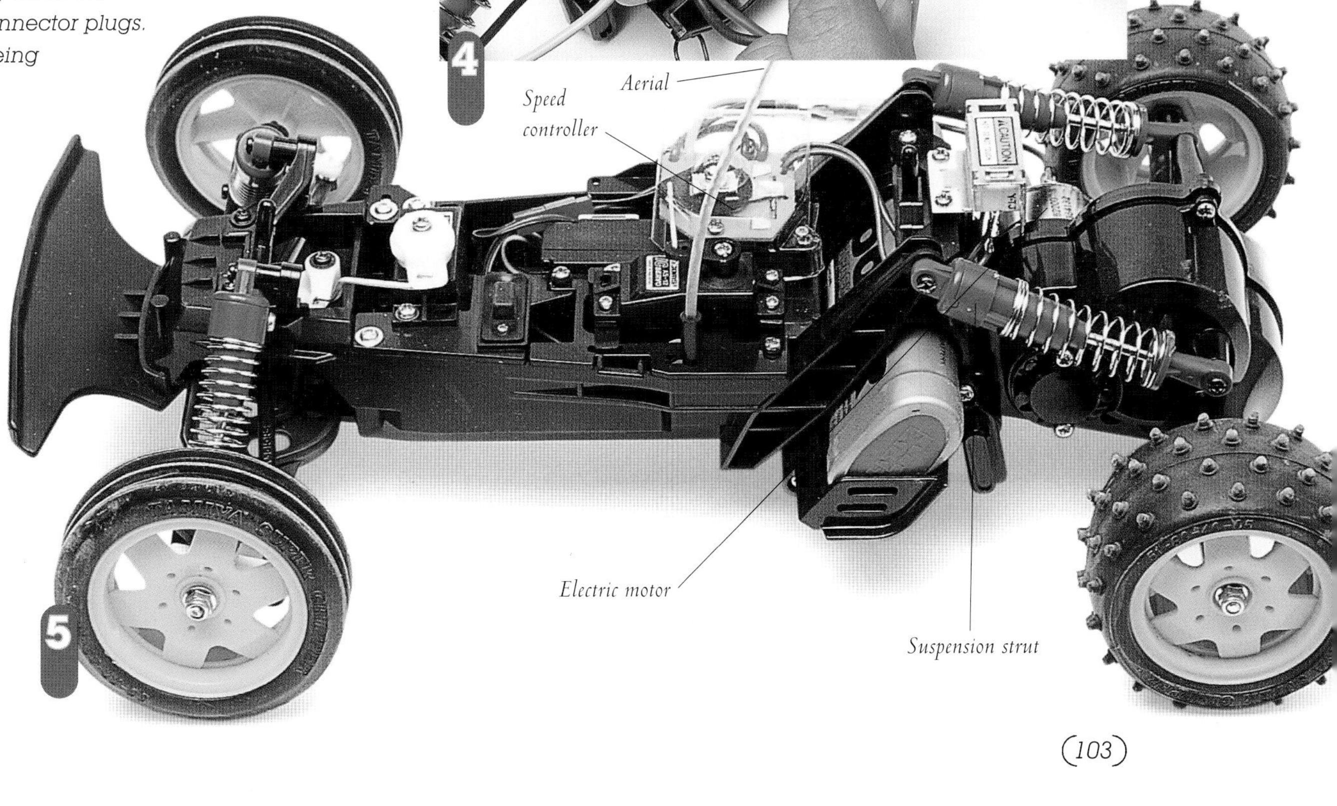

Aerial

Speed controller

Electric motor

Suspension strut

6 Add the decorations—self-adhesive decals—to the body shell, following the kit instructions. The driver is a plain white plastic molding and should be painted. Fit the figure to the Fighter Buggy by screwing through a hole into the helmet top.

7 Before you put the body on, test all the radio-control functions of the model in case anything needs adjusting. Then affix the body by inserting clips into the kingposts. This enables you to make some degree of adjustment to the body height. For cross-country running, for example, you may want to set the body higher, and you can do this by putting the clips higher up on the kingposts.

The completed Fighter Buggy RX has a lively off-road performance.

BUGGY OPTIONS
A range of "hot-up" options for this model are available, including slick tires for racing on smooth tarmac surfaces. In addition, the front suspension is adjustable. If you are operating on a smooth surface, use the inner socket for the damper arms. On rougher terrain, use the outer socket.

SEE ALSO
Basics of radio control, pp.10–11
Driving, p.97
Dos and don'ts of model driving, p.97
Introduction to radio-controlled cars, pp.90–101
Racing, pp.98–101
Radio-control for cars, pp.94–95

This quality model has superb performance, simulating the 4WD characteristics of the life-size Audi A4 car.

TOP OF THE RANGE 4WD CAR

- *1:10 scale model of the Audi A4*
- **Precision engineering**
- **Superb roadholding qualities**

This recent Audi A4 model kit is an example of the new advances that are made in model car technology each year. This model is a clever scale reproduction of the popular Audi A4. Built to 1:10 scale and with four-wheel drive, it looks just like the real Audi A4 and has high performance and good roadholding. Compared with the 2WD Fighter Buggy RX (see pages 102–5), this kit looks very complex. If you are new to radio control and don't have previous kit-building experience, it is a good idea to try the Fighter Buggy first.

TOOLS YOU WILL NEED
Side cutters
Box wrench
Allen key
Screwdrivers
Craft knife
E-ring tool
Small scissors

OTHER REQUIREMENTS
Batteries and battery charger as
  specified for model
Damper oil
Grease
Control equipment as specified

Assembly and control
As is usual with this type of model, the control equipment is not included in the kit but must be purchased separately. Your dealer will have the servos, receiver, and transmitter required to operate the car. The model has differential gearboxes front and rear which need to be assembled. You will also need to paint the clear molded body. The kit also comes with a set of colorful competition-car decals.

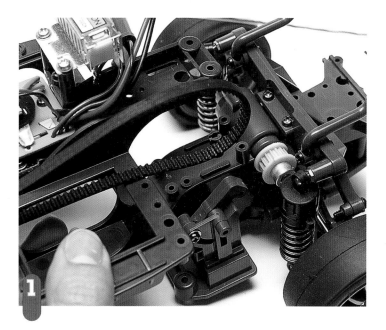

1 *Construct the chassis following the stages outlined in the instructions. Add the rear gearbox and suspension to the chassis, as shown here. The tension pulley for the rubber drive band can be seen in place.*

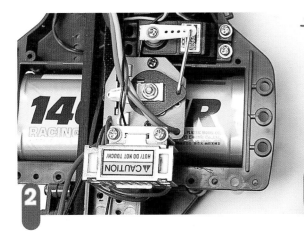

2 Place the front steering servo, the receiver, and the electronic speed control units in position, and then fit the ni-cad battery transversely to the bottom of the chassis. Secure it in place with end plates and clips.

3 The electric motor fits neatly at the front, inside the chassis pan and is held in place by the front suspension unit which is here being screwed into place. Note how the chassis pan acts as a shield for all the working parts which are attached above it.

4 When all the wiring is complete and the receiver and servos are in place, test the steering from the transmitter. The wheels should move in response to the stick, but you may need to adjust the steering so that the wheels point straight ahead when the stick is centered.

5 A critical stage in construction is the fitting and adjusting of the front suspension unit and the fitting of the rubber drive band at the correct tension. Here, the front suspension is being checked for fit.

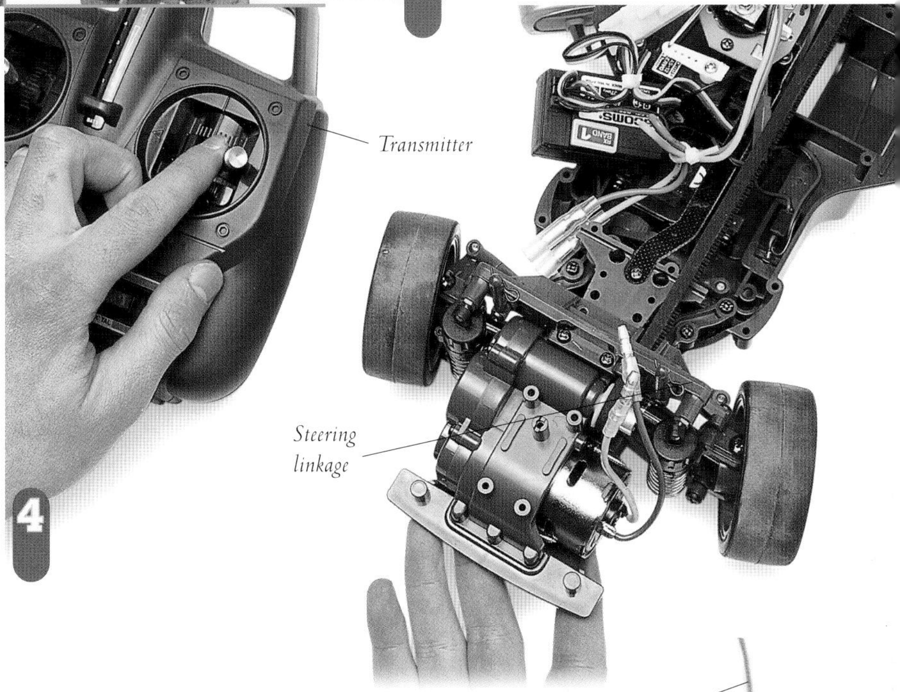

Transmitter

Steering linkage

Aerial

Steering servo

Chassis pan

Rubber drive band

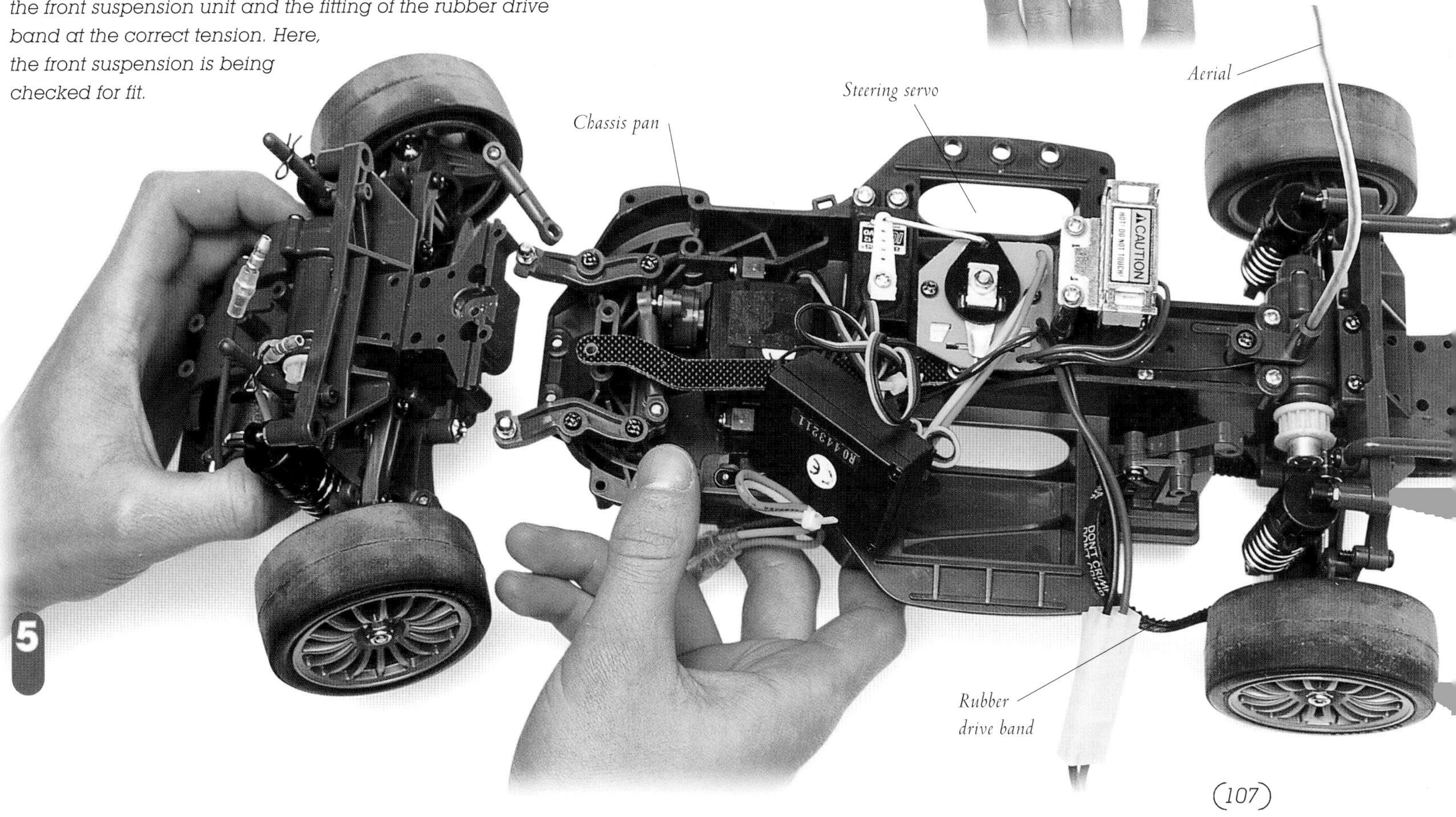

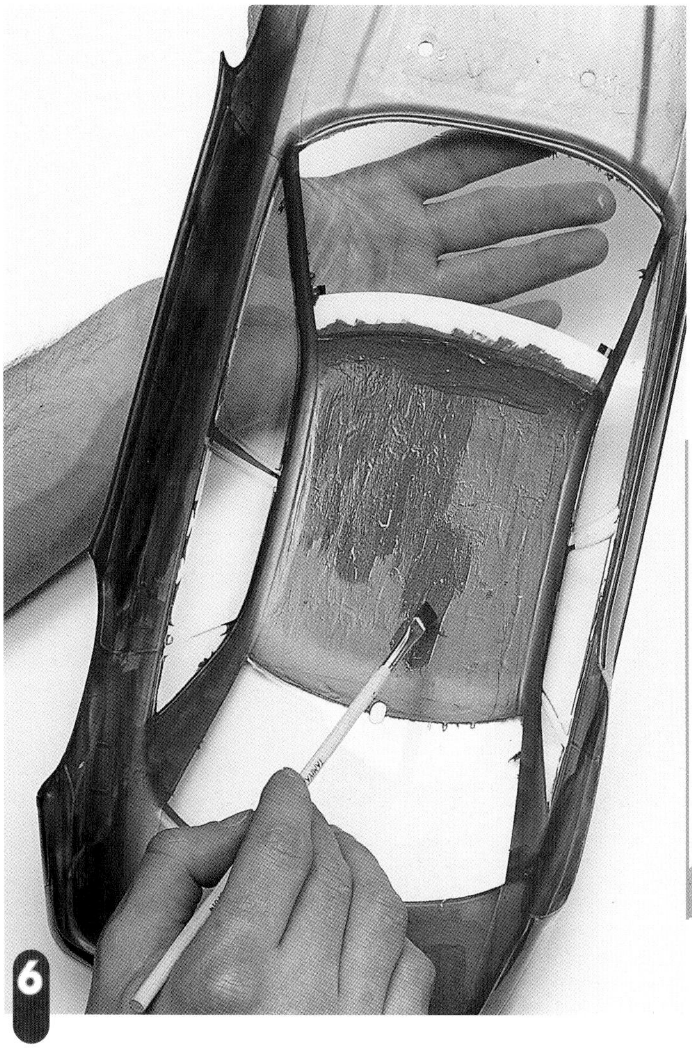

6 When the chassis is complete, the clear molded body can be painted. You will need more than one coat; here, only the first coat has been applied.

7 As is common with most radio-controlled cars, the body is secured with quick-release clips on kingposts. There are usually several positions available, which allow you to set the body high or low depending on whether you intend to run the car on a smooth surface or off-road. Check that the clips fit correctly, as shown here.

The completed model, decorated with typical rally-car markings looks impressive. For a plainer finish, use only a few of the decals supplied with the kit.

THE CAR AERIAL
Important on all radio-controlled model cars is the aerial, which usually comes out through a hole in a window or the body. Always make sure that the aerial is positioned through the hole before fixing the body in place. If it is trapped or snagged inside the body you will be unable to control the car.

SEE ALSO
Basics of radio control, pp.10–11
Dos and don'ts of model driving, p.97
Introduction to radio-controlled cars, pp.90–101
Painting and finishing, pp.22–27
Racing, pp.98–101
Radio control for cars, pp.94–95

This shaft-driven 4WD model is a replica of the Ford Cosworth Repsol Escort rally car. With its dramatic colors and high performance, it is a very satisfying project for anyone with some kit-building experience.

Damper grease

Suspension arm parts

Coil spring

Washers

Damper parts

Side cutters

ADVANCED ELECTRIC CAR

- ### True scale model
- ### Nimble and sporty performance
- ### Precision components

The Ford Cosworth Repsol Escort, manufactured by Tamiya, is an accurate 1:10 scale replica of a successful rally car. All the authentic sponsor markings are included with the kit and the Repsol team colors can be applied to make a spectacular model. The car makes use of many of the latest developments in model technology, including shaft-driven 4WD which gives it superb roadholding and performance. Although it is not recommended as a first choice for beginners, it is much easier to assemble than some advanced kits due to its excellent design, careful engineering, and precision of the parts. It is ideal for a modeler with some simple kit-building experience who wants to tackle something a little more advanced.

The kit contains everything you need to build the rally car, except for the control equipment, which is available separately from your local dealer. As is usual with Tamiya models, the kit also includes a few important tools such as an Allen key.

TOOLS YOU WILL NEED

Side cutters
Box wrench
Allen key
Screwdrivers
Craft knife
E-ring tool
Small scissors

OTHER REQUIREMENTS

Batteries and battery charger as specified for model
Damper oil
Grease
Control equipment as specified

Options

If you wish to enhance the performance of your model, several "hot-up" accessories are available, such as radial racing tires and alternative pinion gear sets with a choice of settings.

1 *Assemble the suspension damper cylinder and insert the grease, taking care to avoid spillage. Put the springs over each damper arm—there are four dampers, one for each wheel—and clip them in place under tension.*

2 *Fully assemble the front suspension and the front differential. Lubricate the suspension arms and linkage using the tube of grease supplied with the kit, and then assemble them together.*

3 *Finally, join the front suspension arms to the differential housing. It essential that all parts move freely without sticking or jamming in any way. Assemble the rear suspension in the same way as you attached the front suspension, fixing the rear suspension units to the rear differential.*

4 *Next screw the front and rear differential/suspension subassemblies onto the chassis pan.*

5 *Take the steering servo from the control set and attach the horn that will take the linkage rod. Screw the steering servo into position on the chassis pan.*

6 *Identify the steering rods and linkage from the kit components and set them out in order of assembly.*

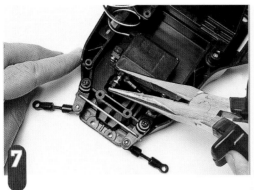

7 *Assemble the steering linkage and associated track rods on the chassis pan.*

8 *Affix the front differential/suspension subassembly to the chassis pan and attach the steering rod to the servo.*

9 *Fit the tires on the wheel hubs and then attach the wheels to the stub axles using the box wrench.*

10 *Fit the receiver on top of the chassis pan and connect it to the electric motor. Now install the electronic speed unit above the receiver. Connect it to the receiver using the fitted wires.*

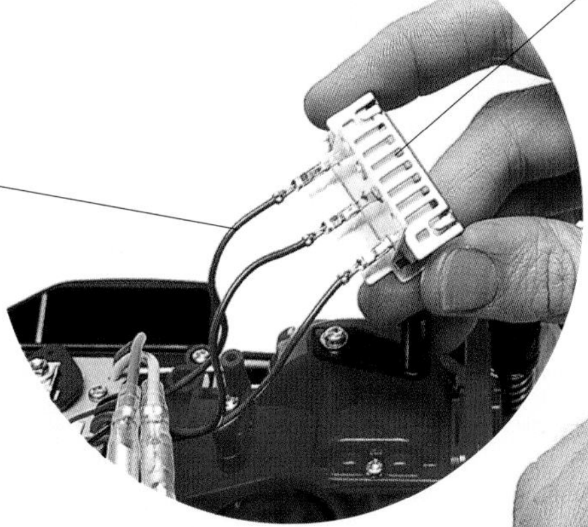

Electronic speed controller

Leads to motors

Shaft drive

11 The shaft drives to the front and rear wheels can be seen here. Also visible in this view of the chassis is the receiver, servo, and the electronic speed control unit. Refer to the instruction sheet for the precise positioning of these items.

12 Insert the ni-cad battery pack in the chassis pan and connect it to the control units. Put batteries in the transmitter and test the control equipment. Check that the wheels move correctly in response to the control stick, as shown here. You may need to adjust the steering so that the wheels point straight ahead when the stick is centered.

13 Now that the chassis work is complete, turn your attention to the body shell. Since this is a clear ABS vac-form molding, you will need to trim away any unwanted excess plastic from the wheel arches and base. When all the edges are smooth, mask off the window areas and paint the shell on the inside with a brush or spray paint. More than one coat may be necessary.

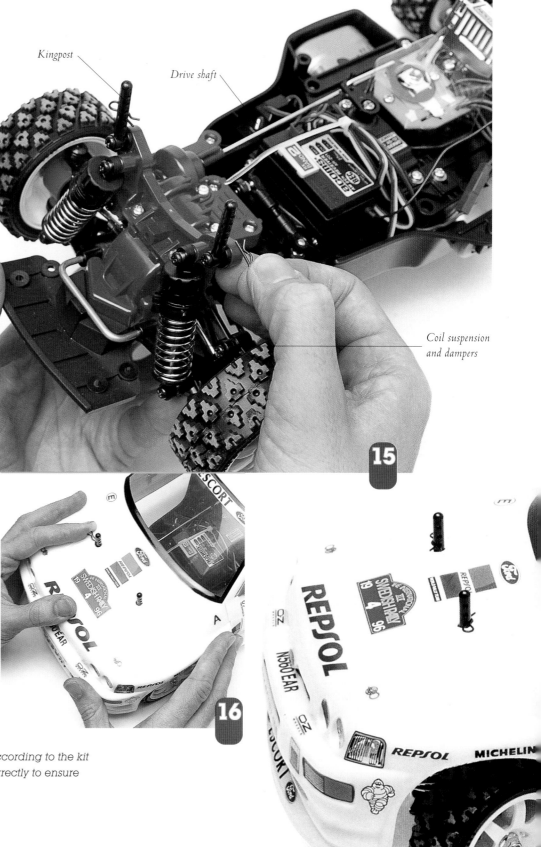

Kingpost

Drive shaft

Coil suspension
and dampers

14 Cut out the decals from
the sheet supplied and fit
them to the painted body
shell according to the
diagram in the kit
instructions. Although this is
an easy task, taking care
when you line them up and
apply them will lead to a
neater look.

15 Now fit the kingposts which hold
the body shell. Identify the clips that
hold the body securely on the
kingposts.

16 Fit the body shell over the kingposts
and secure it in place with the clips.
You can vary the height of the body
over the suspension by fitting the clips
into the higher or lower holes. The body
is normally set higher for off-road
running. Now fit the all-important aerial according to the kit
instructions. It is essential that it is fitted correctly to ensure
successful radio control.

Your colorful model is now ready to drive on its first rally.

THE FINISHED MODEL
Built and finished exactly in accordance with the kit instructions, this completed Ford Cosworth Repsol Escort model is a fine example of the quality engineering and excellent design of many modern kits.

SEE ALSO
Basics of radio control, pp.10–11
Dos and don'ts of model driving, p.97
Introduction to radio-controlled cars, pp.90–101
Painting and finishing, pp.22–27
Racing, pp.98–101
Radio control for cars, pp.94–95

This superb model has an internal combustion engine, giving it an exciting performance and authentic engine noise.

ADVANCED GAS CAR

- **High-precision engineering**
- **Shaft-driven 4WD**
- **Advanced suspension design**
- **Simple screw assembly**

Although most radio-controlled car models, whether made-up or in kit form, are electric drive, some models, such as this Tamiya Honda NS-X, have an internal combustion engine. These gas models seem just like real cars, with their fuel smell and engine exhaust. They are operated by radio control in the same way as electric-drive models, but there is the added thrill of the engine sound and rugged performance. Starting up and running gas models can be far more exciting than running an electric car, but they do have several disadvantages. The fuel for a gas model is an extra cost and sometimes it can be difficult to find a supplier. In addition, the engine noise may irritate your neighbors if you run a gas model in your driveway or backyard. Although modern models are equipped with more effective mufflers than in the past, they do not entirely eliminate the engine noise, and this may restrict the areas where you can operate the model.

TOOLS YOU WILL NEED

Side cutters
Box wrench (supplied with kit)
Allen key
Screwdrivers
Craft knife
E-ring tool
Small scissors
Mini-drill

OTHER REQUIREMENTS

Batteries and battery charger as
  specified for model
Damper oil
Grease
Control equipment as specified

Specifications

The Tamiya Honda NS-X is built to 1:8 scale, making it a satisfyingly large model, about 22in (550mm) long, 9.5in (240mm) wide and with a 12in (300mm) wheelbase. Like most models manufactured by ▶

1 *First assemble the front steering stub axles and suspension, as shown. Next, the oil-filled dampers and associated coil springs are added. Although it looks complicated, assembly is quite straightforward if you follow the screw-assembly sequence, starting with the pre-formed metal chassis pan components which sandwich a center beam "backbone" between them to give immense strength.*

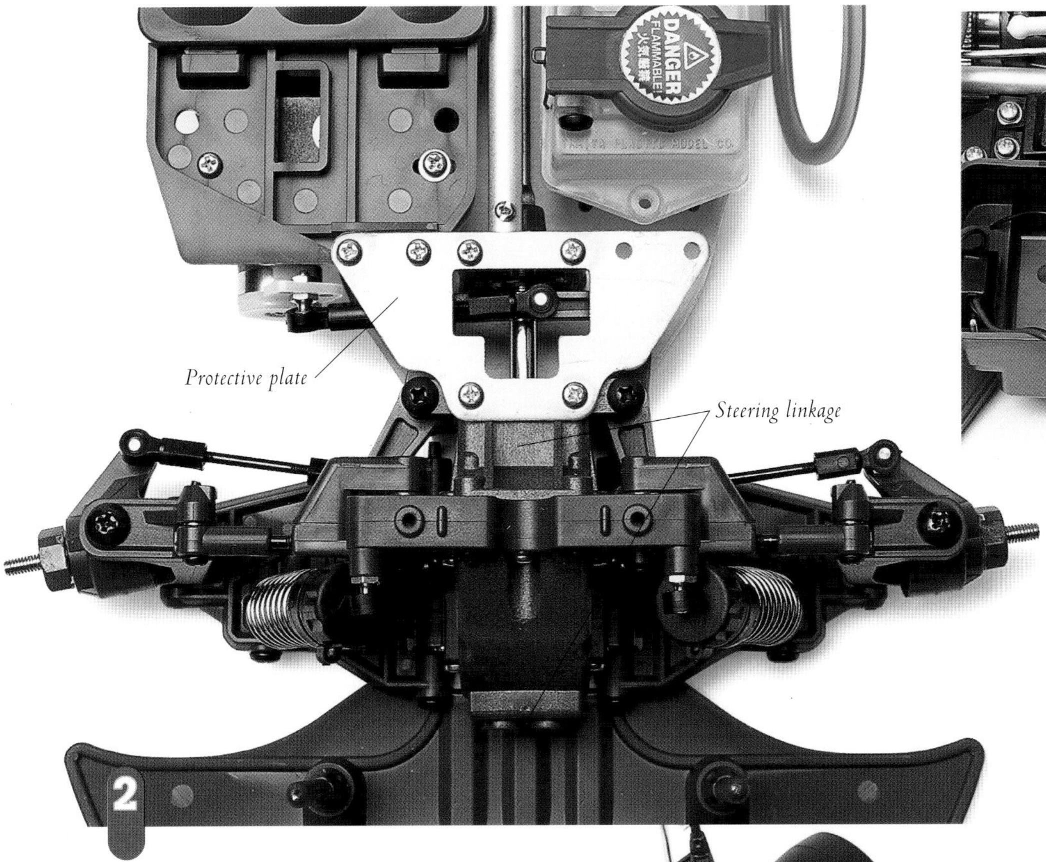

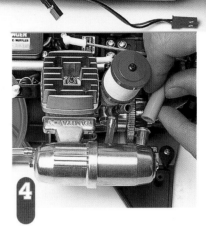

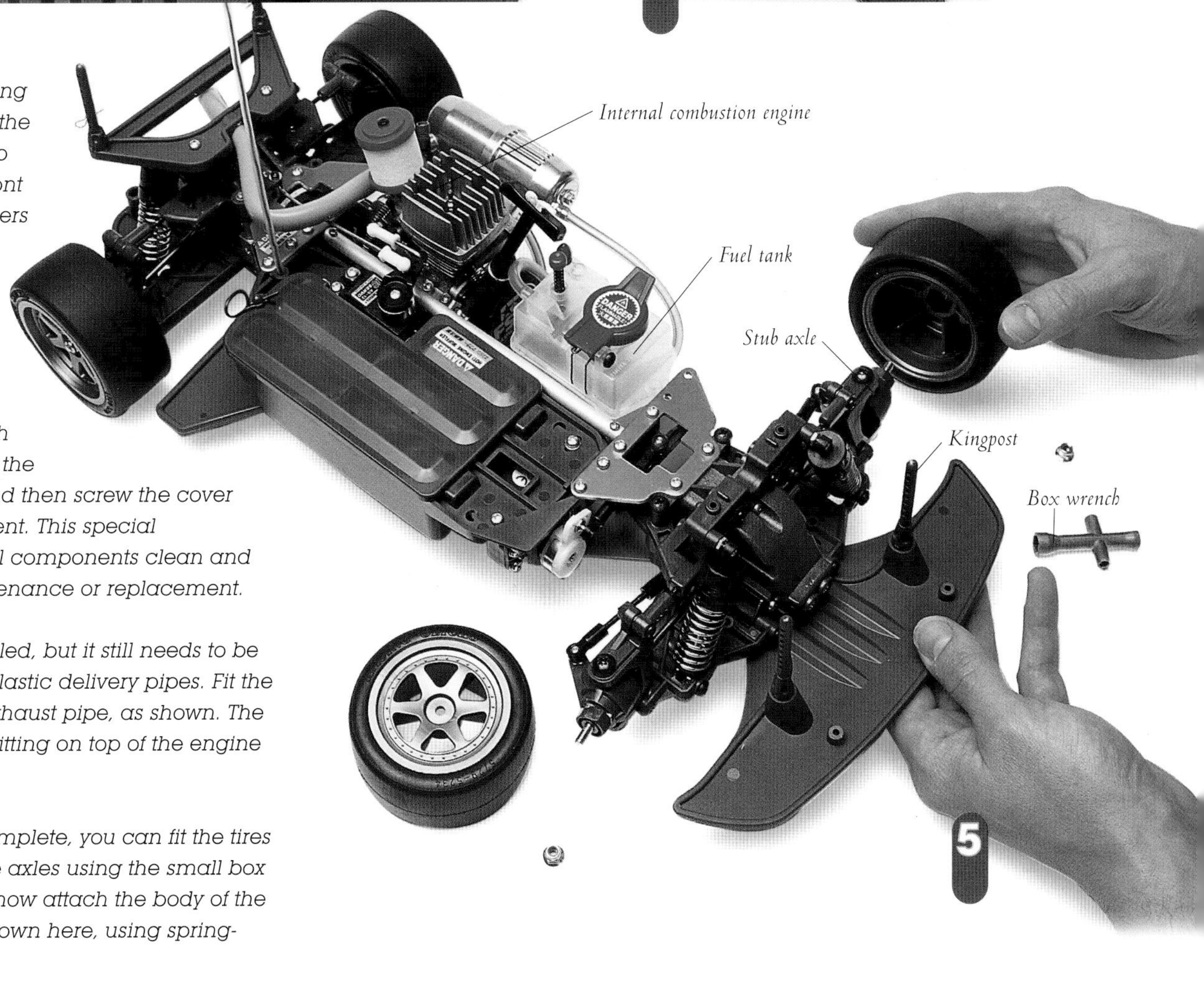

Protective plate

Steering linkage

2

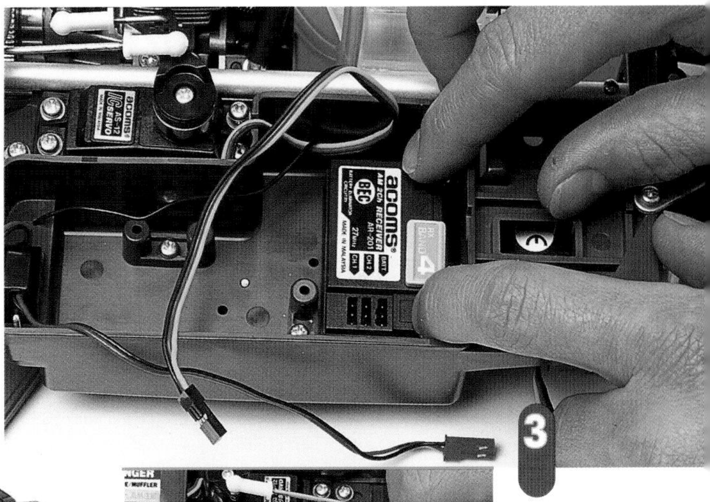

Internal combustion engine

Fuel tank

Stub axle

Kingpost

Box wrench

5

2 *Assemble the linkage from the steering servo to the track rods and then screw the metal protective plate in the center into place. Shown here is the completed front suspension with the oil-damped cylinders and coil springs in place. Underneath the protective plate is the steering linkage and above the protective plate is the compartment holding the control equipment.*

3 *Install the receiver unit into the tough molded plastic compartment. Wire up the throttle servo and the steering servo and then screw the cover plate into place to seal the compartment. This special compartment helps to keep the control components clean and also makes them accessible for maintenance or replacement.*

4 *The IC engine comes ready-assembled, but it still needs to be installed, as do the fuel tank and the plastic delivery pipes. Fit the exhaust muffler and then attach the exhaust pipe, as shown. The motor is started by a pull handle. The fitting on top of the engine is an air filter.*

5 *Now that the chassis assembly is complete, you can fit the tires on the wheels. Attach the wheels to the axles using the small box wrench supplied with the kit. You can now attach the body of the finished model to the four kingposts shown here, using spring-clips to hold it in place.*

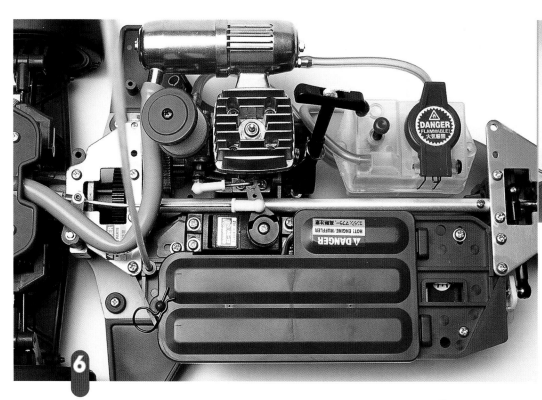

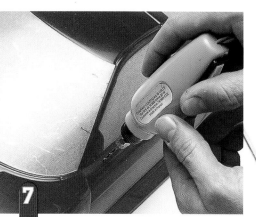

6 *From above, the completed chassis looks complex, but if you follow the instructions step by step, the car can be built with nothing more than a screwdriver, an Allen key, and a box wrench. Note here the covered radio control equipment compartment that hides all the electronic parts.*

7 *Test the completed chassis for radio control functions and for the internal combustion engine function by starting it up with a minimum of fuel. Make any final adjustments—you may need to adjust the steering— and set it aside while you paint and decorate the body. First trim and clean the molding. There are several options for body finishing; this Honda was sprayed on the inside with an airbrush to give a professional finish. Here, the windows can be seen masked while a hole is drilled in the window pillar of the painted body for attaching the rearview mirrors.*

Tamiya, the engineering standard of the chassis, drive, and motor is very high. The specifications for the model make it sound like a real car: it has double wishbone suspension; oil-filled suspension dampers; front gearbox; aluminum upper and Duralumin lower chassis sections; an aluminum center beam, or "backbone"; IC engine; exhaust muffler; air cleaner; and flywheel. The IC engine supplied with the chassis is the OPS VR-15S, though some kits have a variation of this, the FS-15LT. A special feature of this model is the compact, specially sealed compartment carrying the servos, receiver, and electronic speed controller that protects them from being damaged by dirt or fuel. The chassis is four-wheel drive with shaft drive to the front and rear axles, plus full steering. Although the chassis looks complex, it is quite a straightforward screw-assembly job with all parts fitting perfectly. The model needs only basic two-channel radio control with one servo for the steering and one for the throttle.

Customizing your gas model
Tamiya make one standard chassis, TGX Mk 1TS, which is supplied with a variety of different bodies, details, and wheels, according to the model kit. You can also buy the bodies and the chassis separately, so you can make several different cars all based on the same well-proven highly-engineered chassis.

The finished model is an attractive car with excellent performance and road-holding, due to the shaft-driven 4WD feature. Although it looks and performs like an advanced model, it can be assembled by anyone with minimum experience of building radio-controlled models.

SEE ALSO

Competition rules and racing, pp.98–101
Radio control for cars, pp.94–95
Internal combustion engines, p.93
Introduction to cars, pp.90–91
Painting and finishing, pp.22–25

The 18-wheel truck/trailer combination, known as a "big rig," or simply "rig," captured the imagination of modelers many years ago. These trucks are truly kings of the road: big, colorful, dramatic, and of astonishing variety.

Several different semi-trailers are available in kit form for radio-controlled tractors to haul; shown here is a flatbed version attached to the Tamiya Ford Aeromax tractor.

INTRODUCTION TO RADIO-CONTROLLED TRUCKS

Big rigs are powerful machines, packing huge horsepowers to move their heavy cargos down the highway, usually covering vast distances. There are distinctions between American and European brands of truck, but the leading types are known worldwide no matter where they originate. Every truck enthusiast is familiar with such names as Ford, Mack, Kenworth, Peterbilt, Volvo, Mercedes-Benz, MAN, and Renault, and many of these are available in kit form.

RADIO-CONTROLLED RIGS

Recently there has been increasing interest in radio-controlled models of big rigs. Several types of model trucks and accessories are available, including kits and ready-to-run models. The latter are mainly intended for young enthusiasts but they are of interest to those who want to sample a radio-controlled truck before starting a complex kit. Radio-controlled trucks are particularly exciting to build because the huge variety of intricate fittings, decorative details, and color schemes found on real rigs is accurately reproduced on model trucks. For example, the distinctive fleet markings and colors used by major truck companies can all be found. In addition, you can decorate your model truck with many of the optional features and fittings found on real rigs. For example, additional fuel and water tanks and big animal guards that are used on trucks that cover routes in certain areas, such as the American or Australian deserts, can all be added to your model to give it extra rugged character.

Radio-controlled trucks operate on the same well-proven principles as radio-controlled cars. However, trucks are much bigger, so there is relatively more detail, usually more power,

and a lot more genuine miniature engineering involved in the construction. If you are interested in sheer engineering quality as well as the principle of radio control, you will get double pleasure from building a radio-controlled truck kit. You will end up with an impressive model that is a striking display piece even if you never operate it!

OPERATING A RADIO-CONTROLLED KIT

Operating a truck by radio control is a different experience from handling a model car. There is none of the skittishness of lightweight model cars, since the ponderous weight of a model truck simulates the performance characteristics of a real truck

Most of the features of a real truck are reproduced in miniature on radio-controlled model trucks. The complete rear transmission of the Tamiya 1850L truck looks very similar to the real thing.

quite closely. Most models have multispeed transmission, equivalent to gear shifting, full steering, and the handling problems you would need to master if you were driving a real rig. You may have wondered what happens when you back a truck/trailer combination. Which way will the trailer move? With a radio-controlled model you can find out the practical way, but without the risk of doing any real damage if you get it wrong.

MODEL MAKERS

The most widely available truck kits are those made by Tamiya, a leading manufacturer with good world distribution. Tamiya kits feature excellent design and engineering and come with highly detailed instruction books.

In addition to tractor units, Tamiya also make the trailers for the trucks to haul, including box van, flatbed, and tanker types. A wide range of accessory kits are available, enabling you to add scale realistic features to your model. You can choose from a wealth of accessories, including lighting units for both tractor trucks and trailers, oil shock absorbers, animal guards, spoilers, and support legs.

The other major manufacturer of model truck kits is the German company Wedico. Wedico produces model trucks and accessories only, and so has a much bigger range than Tamiya. Wedico models are built to 1:16 scale, which does not match the realistic Tamiya scale, although visually there is very little difference. The Wedico catalog features many drive train and tuning options as well as an enormous array of trucks, including racing trucks. There is even a radio-controlled ferry, which can carry a full rig, so it would be possible for you to drive the rig onto the ferry, carry it across to the other side of the pond, and then drive it off the ferry at the other end, all by using radio control.

However, Wedico kits and associated accessories are not widely distributed, and can generally be found in specialist outlets only.

This Tamiya Ford Aeromax tractor hauling a model of a 40ft (12m) van trailer, captures the thrilling atmosphere of American highways. Scale is 1:14.

TRUCK CONTROL INSTALLATIONS

Truck kits usually include the following control items:

Ni-cad rechargable battery

This provides power supply for the whole vehicle.

Switch panel

Because of the size of model trucks it is usually possible to locate a switch panel on the chassis sides or below the cab. The switches control the battery pack (on/off), radio remote control receiver (on/off), lighting system (on/off), if fitted, and sound system (on/off), if fitted.

Electronic speed controller

This takes orders from the receiver and regulates the speed of the electric drive motor in either direction. You can connect brake lights to this with a suitable circuit so that they come on automatically as the truck stops.

Lighting PCB (printed circuit board)

More advanced kits provide a PCB for all lighting connections.

Turn indicator switch

In more advanced models you can wire a switch to the front wheels so that the appropriate turning lights blink as the wheels are turned left or right.

Sound system

Sound systems are available for two-tone airhorn and diesel engine effects and these can be connected to the switches or the transmission (by using a suitable module) to run while the model is operating. The sound system module made by Wedico is small enough to fit inside a dummy fuel truck on the chassis side or in the back of the truck's cab.

Lighting connection

Most kits include wiring and sockets so that the lighting system can be connected to the semi-trailer or a full trailer, operating rear, brake, and turn lights, as on a real vehicle.

Controls

The most basic requirement for a radio-controlled truck is two servos and two-channel installation. One servo operates the steering and the other the forward/stop/reverse function of the motor. However, most advanced models have multispeed transmission for the motor and need four-channel control to accommodate this and any other function. Some of the advanced models made by Wedico have additional operating features, such as a winch and crane (for a wrecker truck), a hydraulic ram (for a tipper or dumper truck), and remote control (additional to the switch panel) for the lighting system. A multifunction switch is used for these auxiliary tasks. For the most complex models with all these working features, seven-channel radio control is necessary.

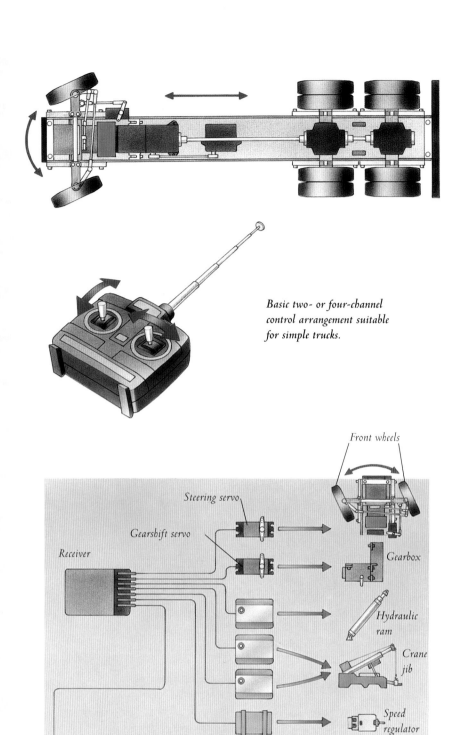

Basic two- or four-channel control arrangement suitable for simple trucks.

Complex radio-controlled models fitted with extra working features, such as lights and horns, may need multifunction switches and extra control channels.

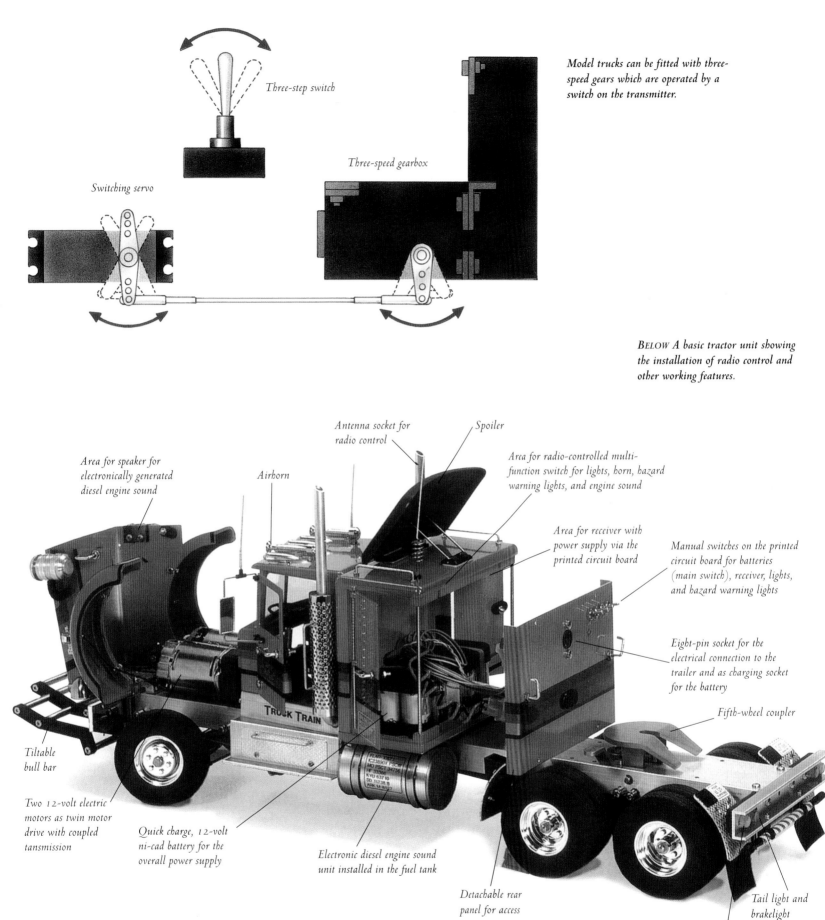

Three-step switch

Three-speed gearbox

Switching servo

Model trucks can be fitted with three-speed gears which are operated by a switch on the transmitter.

BELOW A basic tractor unit showing the installation of radio control and other working features.

Antenna socket for radio control

Spoiler

Area for radio-controlled multi-function switch for lights, horn, hazard warning lights, and engine sound

Area for speaker for electronically generated diesel engine sound

Airhorn

Area for receiver with power supply via the printed circuit board

Manual switches on the printed circuit board for batteries (main switch), receiver, lights, and hazard warning lights

Eight-pin socket for the electrical connection to the trailer and as charging socket for the battery

Fifth-wheel coupler

Tiltable bull bar

Two 12-volt electric motors as twin motor drive with coupled tansmission

Quick charge, 12-volt ni-cad battery for the overall power supply

Electronic diesel engine sound unit installed in the fuel tank

Detachable rear panel for access

Tail light and brakelight

Turn indicator light

MATERIALS AND EQUIPMENT

Although plastic molded parts are used in both Tamiya and Wedico kits, almost all the important "strength" components are made of metal—mostly aluminum or die-castings. In Wedico kits, metal is used for chassis components only. Truck models require considerable screw and nut-and-bolt assembly, so you will need a good selection of small and medium screwdrivers, in addition to basic modeling tools. You will also need a small power drill with ³⁄₃₂in (2.5mm), ⅛in (3mm), ⁵⁄₃₂in (4mm), and ³⁄₁₆in (5mm) drills.

SETTINGS AND ACCESSORIES

Because of their long history of involvement with radio-controlled trucks, Wedico produce many useful and interesting accessories. You can choose from a wide range of options, including operating wrecker and tipper bodies, and many types of semi-trailers and low-loaders. In addition, every conceivable fitting, from decorative wheel hubs and working hydraulic lifts to a variety of cargo and trucker figures, can be found. Many colored decals featuring company markings and registrations are also available.

Other Wedico accessories can be used to make the operation of your truck seem even more realistic. You can make a miniature highway with large indoor layouts or use special markers to turn your yard into a scale-size four-lane highway. To complete this highway world in miniature you can even build a realistic truck stop with snack and refueling facilities, all available from the Wedico catalog. Try testing your driving and maneuvering skills by marking out obstacle courses and parking areas with special Wedico traffic cones. Once you have mastered these complex skills, you may be interested in taking part in the model truck competitions sometimes held by the major manufacturers, such as Wedico and Tamiya.

TRUCK RACING

A recent development in the world of radio-controlled models is the interest in model-truck racing. In this competitive sport, lightened, streamlined, and highly tuned semi-tractors are raced on circuits normally used for racing cars. The races are often spectacular, and sometimes awesome when a driver loses control. Although truck racing is becoming more popular, there are relatively few racing-truck kits available. The main manufacturer of racing trucks is Wedico. However, if you cannot find a racing truck kit, it is possible to convert stock kits of tractor units by adding plastic card fairings and suitable

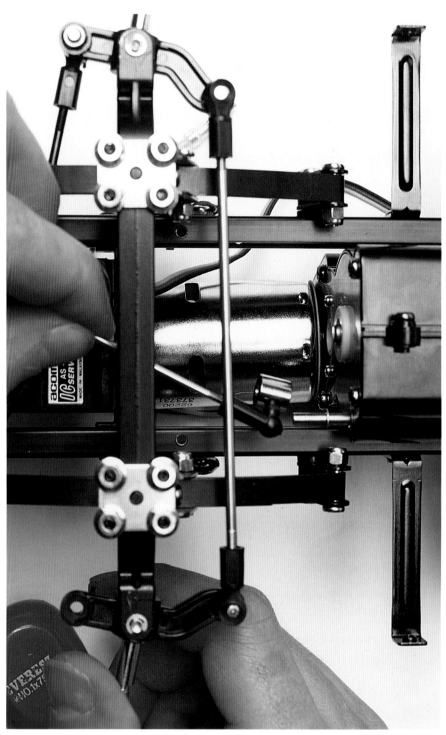

You can learn a lot about practical engineering when you build a large-scale radio-controlled truck. Shown here is the front steering for the Tamiya Aeromax tractor unit, under assembly. The servo that controls the gear shift is being inserted before the wheels are put on the model.

racing decor. As in real truck-racing events, a good deal of skill is required to race a truck, not least because of the high center of gravity and the inherent instability of a unit when operated at top speed around tight bends. One way to improve stability at high speed is to lower the suspension, as on the model racing trucks manufactured by Wedico.

TRUCK GLOSSARY

Conventional
Long-hood truck with engine forward of cab. Known as "normal control" in UK.

Cabover (COE)
Cab positioned over engine, giving flat-fronted (or almost flat-fronted) cab. Known as "forward control" in UK.

Sleeper cab
Cab extended to rear to provide bunk space and overnight facilities for driver.

Prime mover
Short wheelbase tractor unit used to haul articulated semi-trailers. With a ballast body fitted, the prime mover can also tow full trailers.

Semi-trailer
Trailer with rear wheels or multi-axle bogie and front coupling for connection to prime mover. A semi-trailer and prime mover together is known as a semi. There are various types of semi-trailers: tankers, flatbeds, dropsides, vans, containers (known as swap bodies in Europe), tarp-tops (with tarpaulin on supports), dumpers, tippers, and low-loaders.

Truck
Rigid chassis vehicle carrying its own cargo body. Can be four-, six-, eight-, or twelve-wheeled.

Full trailer
Trailer with fixed rear axle and steering front bogie attached to truck by a drawbar.

Truck train
A truck hauling a full trailer. Sometimes two, or even more, trailers can be seen in a truck train, though a multiple haul is usually restricted to very long distance trunk routes, such as those in Australia.

Truck bodies
Truck bodies include: dropsides, tarp-tops, flatbeds, vans (sometimes refrigerated), containers (known as swap bodies in Europe), dumpers/tippers, tankers, special duty bodies (such as garbage trucks, gully emptiers, stock carriers, etc.).

Spoiler
Streamlined fairing affixed to cab roof to improve air flow.

Bull bars
Protective bars to deflect stray animals, foliage, etc., mounted on front of truck.

Front spoiler or air dam
Streamlined fairing carried below cab front on some models.

Fifth wheel
Swiveling coupling on tractor to which semi-trailer is attached.

Landing gear
Supports for semi-trailer when detached from tractor. Usually wound up and down on a ratchet; extended when the semi-trailer stands alone, retracted when attached to tractor.

Dolly
Special bogie used on Australian "road trains" to convert semi-trailers to full trailer style so that several can be towed together.

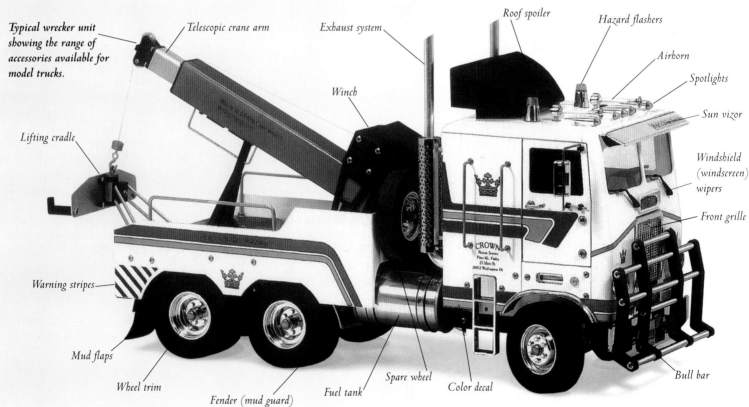

Typical wrecker unit showing the range of accessories available for model trucks.

Telescopic crane arm · Exhaust system · Roof spoiler · Hazard flashers · Airhorn · Spotlights · Sun vizor · Windshield (windscreen) wipers · Front grille · Bull bar · Color decal · Spare wheel · Fuel tank · Fender (mud guard) · Wheel trim · Mud flaps · Warning stripes · Lifting cradle · Winch

The Mercedes-Benz 1850L delivery truck is a model with impressive engineering and superb detailing, fully deserving its title of "king of the road." Although it looks complex, it is a good choice for newcomers to truck models.

KING OF THE ROAD

- *A compact truck model*
- *Impressive detailing*

Colorful and dramatic "Big Rigs," or articulated trucks, can be seen on highways throughout the world. Static scale models of these "kings of the road" have long been popular, but in recent years there has been increasing interest in model trucks with radio control. A range of truck kits is now produced, but the most widely available are 1:14 scale models by Tamiya. The model shown here is the Mercedes-Benz 1850L truck, which is mostly seen on delivery work in towns or on short-haul freight runs. Even though it is the smallest in Tamiya's range, it is still an impressive size when completed: 22.72in (577mm) long and 9.4in (235mm) wide and 11in (275mm) high.

TOOLS YOU WILL NEED
Side cutters
Box wrench
Allen key
Screwdrivers
Craft knife
E-ring tool
Small scissors

OTHER REQUIREMENTS
*Batteries and battery charger as
    specified for model*
Damper oil
Grease
Control equipment as specified

Precision details
In addition to its superb radio-control feature, this truck model has outstanding detailing. Owing to the large size and scale of the model, virtually every feature of the real truck can be replicated on the model, including the ladder-frame chassis.

Performance
With four-channel radio control, servo-controlled three-speed transmission and servo-controlled steering, the Mercedes-Benz 1850L truck model has a superb performance.

1 *First assemble the ladder-frame chassis. Now affix the rubber tires to the wheel hubs and fit the twin rear wheels to the rear axle using the box wrench supplied with the kit.*

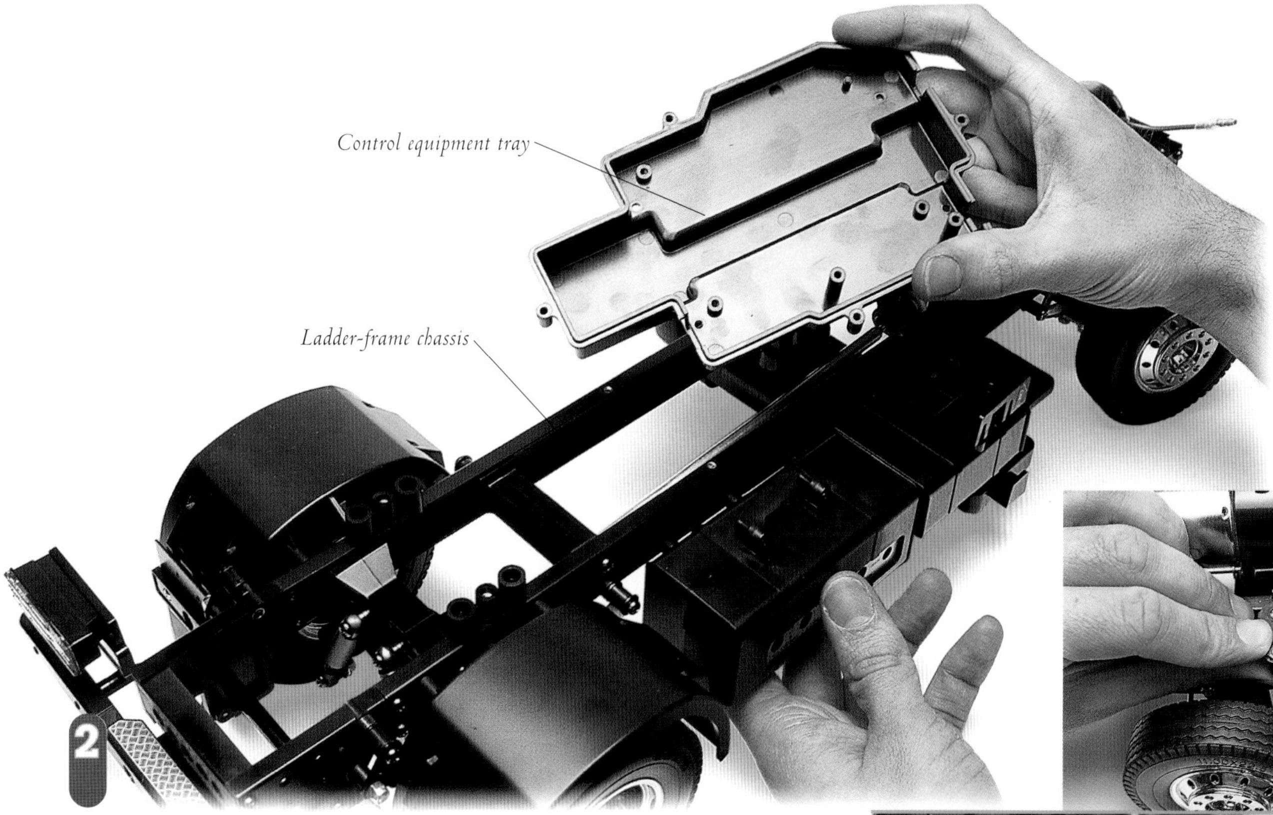

Control equipment tray

Ladder-frame chassis

2

3

2 This detailed model comes with hundreds of parts; identifying and positioning the individual fittings can be time-consuming but is essential for successful completion of the model. Attach the various fuel tanks, battery boxes, and switches to the chassis side members, and then fit the pan that holds the control equipment on top.

3 At the front end, assemble the stub axles, front wheels, and steering linkage, and then attach the operating rod to the steering servo which is located on the chassis side frame. Also visible here is the can-type electric motor wired in at the front of the chassis.

4 The unit that controls the three-speed transmission is fitted in place ahead of the electric motor. You'll find that it fits neatly behind the front grill when the cab is in place.

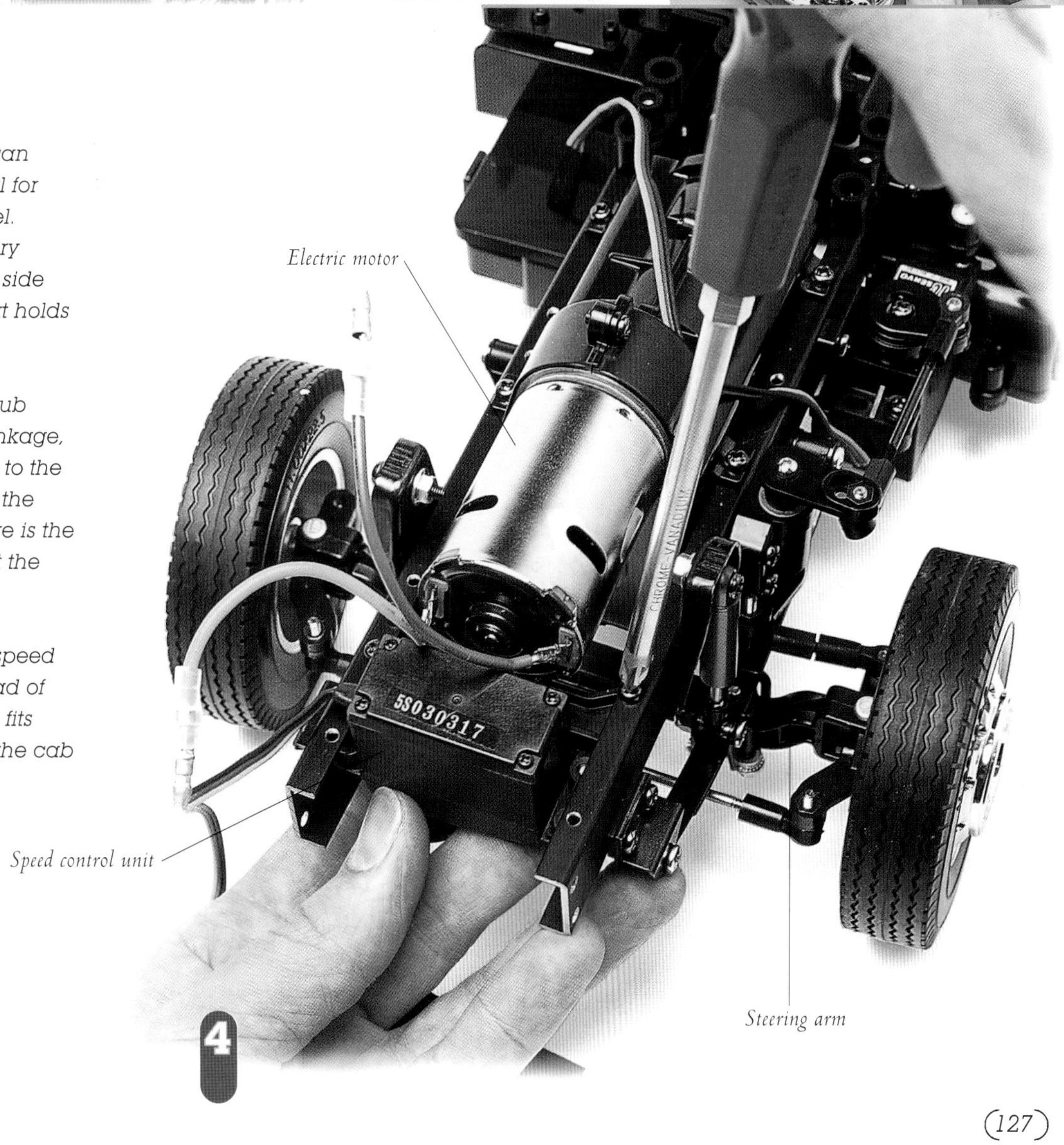

Electric motor

Speed control unit

Steering arm

4

5 *Attach the linkage for the three-speed transmission control servo between the servo and the gearbox. Most assembly work on this large model is by screws or nuts and bolts.*

6 *Bring all the wiring neatly to the chassis side members and join it up to the switches. This convenient and neat arrangement allows the motor and power to be switched on easily from the side of the vehicle.*

7 *Fit the radio control receiver inside the chassis using the double-sided adhesive tape supplied with the kit. Because of the generous size of the model, there is plenty of room for all the control equipment so that the scale realism of the model is not compromised by any unsightly equipment or associated wiring.*

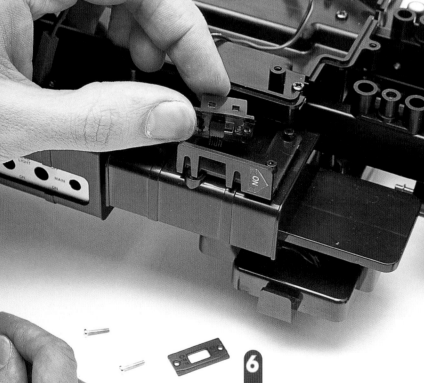

Double-sided tape

Receiver

8

8 Fit the receiver unit in place on the side of the chassis. The pan on top of the chassis holds the ni-cad battery pack that produces the power. Also visible here is the leaf-spring suspension.

9 Now that the chassis is complete and fully fitted out, you can assemble the cab inside and out, and screw it into position on the chassis. The spoiler on the roof is an option which can be omitted if desired.

9

10 Now carefully fit the aerial behind the cab and wire it in. This is one of the most important tasks and must be done correctly to ensure faultless control. Once the batteries are in place you are ready to test the control functions and adjust the steering for accuracy.

11 The final modeling task is to assemble the aluminum body, a pleasing engineering task in its own right, using screws and nuts and bolts. Real hinges are used to assemble the opening rear doors. Once all the assembly work is complete, and the radio control functions have been tested and found to be working correctly, you can paint the model and affix the decals. Though a color scheme and other decorating suggestions are provided with the kit, you can customize your model by using a different finish of your own choice or adding an optional lighting set not provided with the kit.

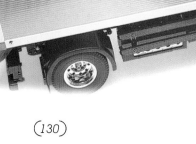

This highly detailed model reproduces virtually all the features of the real truck, including suspension, windshield wipers, rearview mirrors, and opening rear doors. There is also a full cab interior and a set of true-to-type decorations. The body is aluminum and the cab is injection-molded styrene. An optional extra, not shown, is a lighting set that makes all the lights work.

SEE ALSO

Control for trucks, pp.122–123
Electric motors, pp.122–123
Introduction to trucks,
*   pp.120–121*
* Painting and finishing,*
*   pp.22–25*
*   Truck glossary, p.125*

This truly impressive model is a replica of the Ford Aeromax, one of the sleekest transcontinental trucks on the highway. In spite of its size and sophistication, it can be tackled by a modeler with a minimum of experience.

A BIG RIG

- ## The biggest truck kit available
- ## Authentic detailing
- ## Variety of trailer options

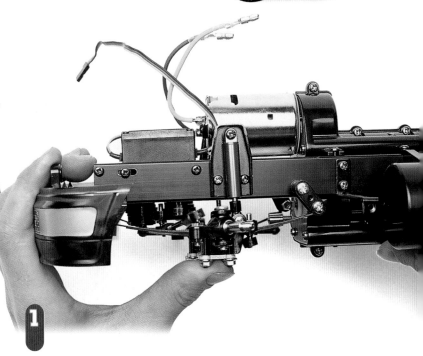

Although several different truck-tractor kits are available, the model featured here, the Ford Aeromax, by Tamiya, is one of the latest and perhaps the most impressive big rigs to be seen on American highways. Built to 1:14 scale like other Tamiya trucks, this model is truly impressive: the tractor unit alone is 25.6in (650mm) long, 12.4in (315mm) high, and 7.6in (190mm) wide. If you attach the van trailer as well, which is 35.44in (900mm) long, you will end up with a very striking big rig, nearly 60in (1.5m) long. In addition, the model has "state of the art" radio control. The truck has an authentic-looking ladder-frame chassis, leaf-spring suspension, an electric can motor, and a three-speed gearbox controlled by servo from the four-channel radio-control system used for this model. The detailing is superb, and with careful painting you will have a very fine and authentic model.

TOOLS YOU WILL NEED

Side cutters
Box wrench
Allen key
Screwdrivers
Craft knife
E-ring tool
Small scissors

OTHER REQUIREMENTS

Batteries and battery charger as
  specified for model
Damper oil
Grease
Control equipment as specified

Building the Big Rig

Some kits look so big and complex that even modelers with some experience feel hesitant about building them. However, in spite of its impressive size, this big rig kit is a pleasing model to assemble and offers few problems. Although there are hundreds of pieces, each part is well engineered and the mechanical parts are made to fine tolerances. ▶

1 Assemble the ladder-frame chassis, the rigid axles, and the front fender. Test the front suspension to make sure it works, as shown here. Install the servo for the speed control (three-speed gearbox) but do not wire it in. Follow the step-by-step instructions carefully here: though it may look simple, there is a considerable amount of assembly work needed to reach this stage.

2 *Assemble the track rods that operate the front steering and the stub axles. Attach and wire in the steering servo. After all the other control elements are fitted, you can wire up the battery pack and test the remote control of the steering to make sure it works.*

3 *An important element is the electronic speed controller: install it on its mounting plate, as shown here. Adjacent is the receiver module and at the front of the chassis are the two servos that control the steering and gear change respectively. These all have to be wired up and connected with the 7.2V ni-cad battery pack that fits in the chassis.*

4 *The switches that enable you to start the motor and operate the lighting system (if fitted) are located on the side of the chassis, neatly concealed by the cab steps. Screw the switch unit in place, as shown here. Identify the exhaust stacks and the aerial support mount. The exhausts are, of course, cosmetic on this electrically-driven model.*

Cab body

5 *The cab unit is a separate construction job. However, all the parts are neatly molded in polystyrene just like a regular plastic kit, making it easy to assemble. Numerous features such as grabrails, rearview mirrors, airhorns, and authentic license markings are included in the kit.*

Trailer options

Tamiya also manufacture a range of trailers for the big rig truck-tractors to haul. These include kits for a flatbed trailer, a van trailer, and a tanker trailer. Building a trailer is a pleasing project in its own right, and if you are particularly interested in truck modeling but have little kit-building experience, it may be a good idea to build a trailer first before moving on to big truck-tractors. The trailer, of course, does not have any radio-control component, but the assembly is a good introduction to making large-scale models and will give you the necessary confidence to tackle the tractor unit.

Truck-handling contests are popular events at model clubs. It takes some practice to master the skill of backing a large trailer into a parking bay. You can practice by marking out the parking spot with chalk.

6 *Attach the body to the chassis and check for any final adjustments. You can paint the screws and fixings to match the rest of the body color. Although this model was painted blue, you can paint your model any color, or even two-tone if you prefer.*

A BIG RIG

THE BIG RIG TRACTOR
The finished Big Rig tractor is shown below, hauling the standard Tamiya van trailer. The Ford Aeromax can haul any of the trailers designed to match the 1:14 scale Tamiya trucks. It is also possible to make up other trailers, either from scratch or from Wedico trailer parts. If you use the optional lighting unit, you will need to include a power connection for the trailer lights.

SEE ALSO
Control for trucks, pp.122–123
Electric motors, pp.122–123
Introduction to trucks,
  pp.120–121
Painting and finishing,
  pp.124–125
Equipment, p.124
Truck glossary, p.125

Tanks and other crawler track vehicles pose a fascinating challenge for the modeler due to their unique characteristics. Other model vehicles included in this category are dozers, excavators of various types, and snow vehicles, since all depend on crawler tracks to spread their weight over the ground surface.

RIGHT This scratch-built Cromwell tank (British, 1944) was made by members of the Model Armoured Group. Built to 1:8 scale, the model is made of plywood. One motor drives each track, and the track shoes are from a child's construction set.

INTRODUCTION TO RADIO-CONTROLLED TANKS

Vehicles equipped with track shoes that grip soft surfaces can cross terrain (including deep snow and ice, in the case of snow tractors) that ordinary wheeled vehicles with standard transmissions cannot tackle without becoming bogged down.

Static and motorized free-running tank models became popular soon after real tanks appeared in World War I. The first radio-controlled model tanks were scratch-built by enthusiasts well before manufacturers started to produce tank kits. A pioneer of radio-controlled model tanks in the early 1970s was the British Model Armoured Group. This group worked to a large scale of 1:8, producing models of an impressive size from plywood. The scale enabled them to use the metal track shoes produced for Meccano (Erector-type) crawler track models.

REPRODUCING TANK MOBILITY

A real tank has impressive mobility, because each track can be driven and controlled independently. This is achieved by a complex gearing system between the engine and the tracks. With this system, the great bulk of the tank can be maneuvered with ease. For example, "neutral turns" can be performed whereby one track runs forward and the other in reverse, both at the same speed, so that the vehicle spins on its own axis without moving forward. Sharp turns and fast backward running are also possible, all of which make the idea of a radio-controlled tank very attractive for model enthusiasts. However, modelers faced the problem of how to replicate the mobility achieved with this complex gearing system. Simpler scratch-built models avoided the problem by adopting the system used on some early real tanks: a separate motor driving each track. Each track could be driven forward or reversed independently by allocating a servo to each motor and using the left and right sticks on the transmitter control box to control the left and right tracks. This works well and some ready-to-run tank models use this method.

ABOVE The cab of the Blizzard snow tractor can be removed for access to the chassis and suspension. Note the wide tracks and front-mounted IC engine.

LEFT The US Army Abrams main battle tank manufactured by Ripmax is an ideal kit for the newcomer to tank modeling. The model performs impressively on cross-country terrain.

BELOW The Ripmax Abrams tank kit has very few parts, making it ideal for beginners. Note the working wire torsion bar axles on the road wheels.

MODEL TANK KITS

Very few radio-controlled model tanks are available, whether ready-to-run or sophisticated high-tech kits. The leading manufacturer of high-quality radio-controlled tank kits is Tamiya. Their kits include the modern German Leopard A4 main battle tank, an anti-aircraft Flakpanzer variant, the Gepard, and the famous World War II King Tiger (Koenigstiger), or Tiger II, which is one of the biggest tanks ever to see service. The scale for these tanks is 1:16. Tank kits are true masterpieces of engineering. The chassis pan is made of

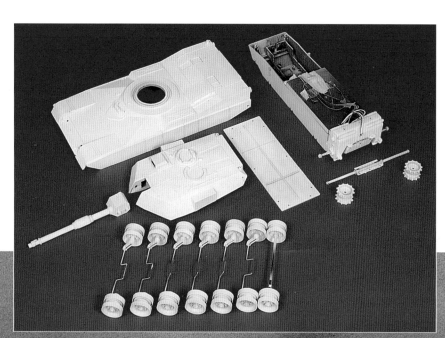

aluminum or duraluminum, and there is a superb twin-clutch transmission, allowing the single electric motor to drive each track independently, just like a real tank. Tracks, drive wheels, and axles are all metal, and proper torsion bar suspension is incorporated as in real tanks. A separate small electric motor drives the turret, just like the real thing. On the Gepard, there is a separate motor and gearing to turn the radar aerial on the turret top. The radio controls allow left, right, forward, and reverse movement, as well as the traversing of the turret and elevation of the main armament. In the King Tiger, there is even a strobe unit to simulate the firing of the gun. However, these tanks are only produced intermittently in limited runs.

GETTING STARTED

Model tanks are quite complex to build, but they are as straightforward as Tamiya car kits in presentation and assembly requirements. At the time of writing, a good tank for the beginner was the American MIAI Abrams kit,

manufactured by Ripmax. Built to 1:16 scale, this tank is a good replica of the famous modern US Army main battle tank. Although it lacks some of the working features found on Tamiya tanks, the Abrams is simple to assemble, has effective torsion bar suspension like a real tank, and runs well over obstacles and rough terrain.

To check availability of current tank models, ask at specialist model car and truck stores, since radio-controlled model tanks are usually categorized with model road vehicles.

OTHER TRACKED VEHICLES

Other radio-controlled tracked vehicles are sometimes available. Kyosho recently produced a handsome radio-controlled model of the Blizzard tractor used in Arctic and Antarctic exploration. This model is controlled in the same way as a tank, but is different from other tracked vehicle replicas as it is powered by an IC engine. Because the track extends well beyond the chassis, final drive to the sprockets is by chain from the drive shaft. With its impressive engineering, this model runs extremely well and will certainly attract a lot of attention.

BELOW LEFT The Blizzard snow tractor taking a soft sandy bank in its stride with its long wide tracks.

BELOW The Blizzard tractor is a fine example of model engineering at its best. Shown here is the final drive by chain to the front sprockets.

Radio-controlled cars are the most popular and accessible type of model. A huge variety of vehicles are available, including F1 racing cars, touring cars, buggies, off-road jeeps, stadium trucks, and big-wheel cars. In general, cars are the easiest for a complete beginner to make but shown here is an array of examples ranging from simple beginner's buggies to complex IC-engine cars.

GALLERY

LEFT *This 1:10 scale replica of the classic Porsche 911 Carrera is a highly detailed Tamiya model operated by basic two-channel radio control. In front of the completed model is the chassis—a 2WD electric unit with drive on the rear wheels, independent suspension, and a sealed gearbox with differential gearing. The steering servo and its linkage can be seen on the top of the chassis. The ni-cad battery fits neatly into a transverse housing below the control equipment.*

LEFT If you are interested in off-road racing with larger scale models, try the Inferno by Kyosho. This shaft-driven 4WD buggy is built to 1:8 scale and has an IC engine. The model is light but strong owing to the extensive use of aluminum for the chassis and upper deck. There is a servo saver to protect the steering servo from damage in rapid maneuvers.

ABOVE The Kyosho 1:10 scale Outrage is based on American stadium trucks used for truck racing. It has a sealed gearbox and 2WD. The model uses the same chassis as an earlier Kyosho buggy, making it easy to build and operate. This model is electric, but for a similar IC-engine model, try the Sandmaster ST.

LEFT The 1:10 scale replica of the McLaren F1-GTR in the Kyosho Pure Ten range is an advanced touring class model. It has an IC engine, 4WD, "double-A" arm suspension front and rear, all mounted on a Duralumin chassis. This particular model also benefits from hot-up options, including "low-profile" racing wheels. It is available in more than one team color, but the Gulf version is the most striking.

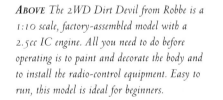

ABOVE The 2WD Dirt Devil from Robbe is a 1:10 scale, factory-assembled model with a 2.5cc IC engine. All you need to do before operating is to paint and decorate the body and to install the radio-control equipment. Easy to run, this model is ideal for beginners.

LEFT *This radio-controlled go-kart, manufactured by Kyosho, is an attractive model in spite of its large size. The chassis uses the same components and mechanics as a model car. Racing model go-karts on specially designed compact circuits is a popular hobby.*

BELOW *This 1:10 scale replica of the famous 1962 Mk 1 MGB sports car is in the Kyosho Pure Ten series. The model is electric-powered and has a mechanical speed controller, giving, in effect, three forward gears and reverse. The realistic gloss finish is achieved by painting the clear polycarbonate body molding from the inside. The molded driver figure and the chrome parts are included with the kit.*

LEFT *This 1:10 scale model of a Ford Escort Cosworth Repsol from the Kyosho Pure Ten range can be operated either with an electric motor chassis or an IC-engine chassis.*

RIGHT *This 3-axle German MAN van comes from the Wedico 1:16 scale range. It has a popular European-type F90 cab, and features authentic Alcoa wheel trims.*

BOATS

Whether you are interested in an exciting high-speed racing boat or a finely detailed harbor craft, a scale-model warship or a graceful yacht, you'll find something to interest you in this section.

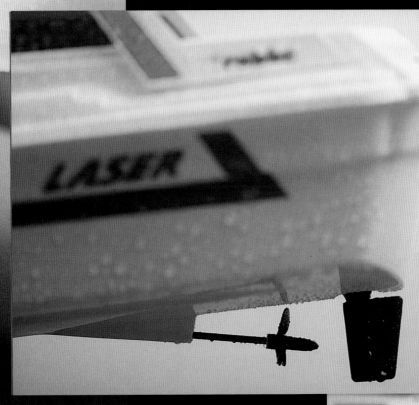

Although powerboats are mostly thought of as motorized racing vessels, the description covers motorized boat and ship models of all kinds.

INTRODUCTION TO POWERBOATS

The wide range of radio-controlled motorized boat kits includes not only the popular racing boats, but also motor cruisers, motor yachts, tugs, fishing boats, harbor service boats, and oceangoing liners. Large warships and submarines are also available.

MODEL-BOAT KITS

Model boats or ships are ideal for radio control. The control equipment fits well in the hollow hulls and has no effect on their stability or "seaworthiness," since it is relatively small and

lightweight compared to the bulk of a big model boat. Elaborate large-scale model kits, including all the necesary wood, materials, and complex plans, are available for the experienced modeler.

At the other extreme, there is a wide selection of models in the ready-to-float (RTF) and almost-ready-to-float (ARTF) category. These models are ideal for beginners as they require little more than fitting the batteries and checking the controls before you put them in the water and set off. In addition, you will get a good idea of how radio control works because you can see it

TOP Large-scale powerboats can look extremely realistic. This is a superb replica of a French fishing boat, complete with crew.

ABOVE Not all powerboat models are made from kits. This is the stern of a superbly detailed radio-controlled British "Battle" class destroyer made entirely from scratch.

already installed in the model. Depending on the kit, you may have to clip in a few details, such as handrails, stump mast, and aerial mast. In the ARTF kits you may need to fit a few larger components in place, too, such as the hull or the cabin tops, after the batteries are installed.

Most other radio-controlled boat kits come with all parts ready to assemble—the molded plastic hull is usually a single piece or in two parts with the deck to be attached to the lower hull. These kits are ideal for those with some modeling experience. In most cases you will have to buy the control equipment separately and fit it yourself. Most of these models are fast-racing powerboats, reflecting the fact that these glamorous vessels appeal to the greatest number of enthusiasts. If you want something a little gentler than off-shore or closed-circuit racers, you can choose from a number of non-racing boat kits, including river cruisers, motor gunboats, fishing boats, and service vessels, such as fire floats and buoy tenders.

BEFORE YOU BUY

Always check a kit carefully before you buy it. Descriptions on boxes or in catalogs can be interpreted in several ways. Although a model may be described as ARTF, it may mean that only the motor and propeller shaft are factory-fitted in a ready-molded hull. You will have to install the control equipment and the hull fittings. Ideal for beginners with some modeling experience, these kits are available from the leading powerboat manufacturers, Kyosho and Robbe.

Also be wary of "beginner's" kits that seem to be made up from easily assembled parts. These can turn out to be more complicated than they look, requiring far more assembly work and skill

RIGHT The Viper, manufactured by Kyosho, is an ideal model for beginners, since it has a factory-fitted motor and speed controller and is easy to build.

BELOW Like many modern powerboat kits, the Kyosho Viper comes with an ABS molded hull.

than kits assembled from finished parts. For example, you may have to use printed paper patterns to cut wood parts. These type of kits are not far removed from scratch-built models, the only difference being that they contain all the materials you need and may have a few molded parts, such as a smokestack or basic ABS-molded hull shell.

BELOW This cutaway of a Kyosho Jetstream 800S shows the typical interior layout of an IC-engine powerboat.

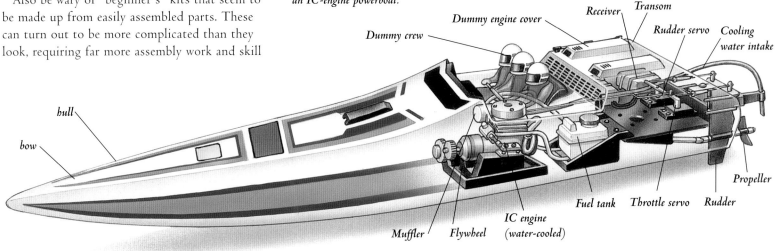

MATERIALS

Kits using wood are still available today, although GRP (fiberglass) molding has largely replaced wood. GRP hulls are very tough yet light and can be repaired quite easily with fiberglass repair sets. The most widely used material today, however, is ABS plastic, which gives tough but light hulls with a smooth finish and surface detail. An interesting recent development is the bonding of wood on an ABS molded hull to give the effect of an all-wood model, but with the toughness and durability of a plastic hull. You may find a few smaller models with a hard plastic polystyrene hull.

Many boats with plastic hulls may have wood parts as well, such as the trays that hold the control equipment inside the hull. Fittings and accessories may also be made in a variety of materials, such as wood, aluminum, hard polystyrene plastic, and wire. Self-adhesive decals are usually used with powerboats, because these stay on well in water.

MOTORS

Although electric motors are the most popular and most practical, IC engines and traditional steam engines are also available.

Electric engines

Most model powerboats use electric motors. Highly efficient and easy to install, these motors come in a range of sizes. They are quiet and do not emit fumes and so are acceptable in public places, such as lakes and ponds. In addition, radio control of the motor is straightforward.

ABOVE The Streamliner by Kyosho is a handsome older-type motorboat. The real wood bonded over its ABS hull gives the model a luxurious handcrafted effect. The transmitter, battery pack, and battery charger for this model are shown in front.

LEFT This close-up view of the Streamliner motorboat shows the electric motor in the forward compartment and the control gear in the back. They are all accessed by removable rear deck hatches.

IC engines

In recent years there has been a revival of interest in powerboat kits with pull-start IC engines. The noise and noxious fumes that were previously the main obstacle to their use have been greatly reduced owing to the introduction of more efficient mufflers. High-performance boats with IC engines can reach spectacular speeds of up to 25 knots or more, compared with a high speed of 20 knots for an electric motor. Modern IC-engine boats are not only dramatic and thrilling, but also well designed. The Kyosho Jetstream 1000, for example, has separate compartments for the motor, fuel, and control gear, and an "easy fill" fuel tank. The boat planes beautifully and stays dry, a key point for all radio-controlled powerboats. All IC-engine boats have water-cooling for the engine, usually through an intake in or behind the rudder post. All the necessary parts are included in the kit, so you do not need to figure out engine cooling for yourself! However, a major disadvantage to IC-engine powerboats is that local ordinances in your area may prevent you from running them in public places. Before you run your IC-engine powerboat, always check whether any restrictions apply in your chosen area.

Steam engines

Another type of motor that you may come across is the steam engine. There are a few companies that produce small marine steam engines in various sizes for model boats. These engines normally come with the necessary gearing and/or shaft. They are most often used for charming scratch-built models made specially for steam propulsion. However, a few radio-controlled boat kits complete with steam engine are available, including an old-time tugboat. These tend to be more expensive than more conventional kits, and may be difficult to find in stock—you may have to order specially.

BELOW Hydrojet high-speed boats are propelled by a waterjet.

BELOW This Kyosho Hydrojet planing at full speed is an example of a high-speed motorboat powered by a waterjet unit.

Hydroplanes and waterjet propulsion

Two other propulsion systems are sometimes used on model powerboats. One old type, known as a hydroplane, has an aero engine with a propeller. This reproduces a type of propulsion by providing either the "push" or the "pull" (depending on the way the engine faces). The rudder is in the usual place and is used for steering in the same way. A recent development on high-speed boats and catamarans is the waterjet propulsion system. The jet units can be used for steering if two are fitted, and even for reversing if the appropriate control unit is used. Boats with only one jet unit use a rudder to steer. So far this system is found mainly in models of actual boats that use them, such as the record-breaking *Atlantic Challenger*.

CONTROL EQUIPMENT FOR POWERBOATS

A simple, RTR powerboat may have only a single channel and one servo operating the rudder. The switch for the electric motor is usually found in the boat's deck or cabin top. All you need to do is put the boat in the water, switch on the motor, push the boat off and then control it with the steering. However, most powerboats built from kits are two-channel. One channel controls the steering. Depending on the kit, the second channel may do any of the following: switch the electric motor on and off; switch the electric motor from stop to forward gear and back to stop and from stop to reverse gear and back to stop; control forward speed of an electric motor, via an electronic speed controller; control the throttle (and therefore, the speed) of an IC engine. What you intend to do with your boat may govern your choice of kit. If you want to race your boat, you will need good speed control. However, if you are more interested in boat maneuvers, a model that offers reverse on the second channel would be more suitable.

More sophisticated models have a third or fourth channel for additional functions. For example, the pump on model fire floats that eject water from a bow monitor just like real fire floats

is controlled by a third channel. Bilge pumps on larger models are controlled in the same way to eject water taken on board. Working derricks on fishing boats and buoy tenders are also controlled by a third channel. In addition, turrets on sophisticated model warships can be trained by the third and fourth channel. Some radio-controlled submarines are so sophisticated that they need four (or more) channels. These control the steering, speed, the hydroplanes to make the boat dive and surface, and a "diving mechanism," or tank placed below the hull which can be flooded to keep the boat weighted below the surface.

ADDITIONAL FEATURES

A useful feature of some of the latest sophisticated powerboat kits is warning lights (LEDs) to indicate when the battery is running low. This is a particularly important feature for a powerboat model, because problems can occur if your on-board battery pack runs out while the boat is in the middle of a lake. Although you are unlikely to lose your boat, you could have a long wait while it drifts ashore. It is a good idea to take a spare fully charged battery pack on a boat outing, and change it if in any doubt about the remaining strength of the battery on board.

ABOVE With its working derrick, the Robbe Norderney harbor service boat is extremely realistic. The derrick is raised and lowered by an extra servo.

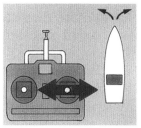

ABOVE The right stick on the transmitter always controls steering, left and right.

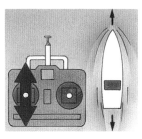

ABOVE The left stick controls direction (forward/reverse) or sometimes engine on/off.

POWERBOAT RACING

If you are interested in racing your radio-controlled powerboat, it is best to join a local club. Check your local newspaper or model-boat magazines to find out about clubs in your area. You can also contact your national association—the American Power Boat Association or the UK Model Power Boat Association—for details of affiliated clubs. The International Model Power Boat Association in the US and Naviga organization in Europe have set up international classes and run international events—current addresses can be found in model-boat magazines.

Racing classes for powerboats are quite diverse as they include electric and IC engines, and many different sizes of models. In addition, the classes set by national organizations do not always match the international classes. For example, Naviga has a Formula 1 (F1) class for racing boats divided according to size, engine power, and type, and an F2 class strictly for scale models. These are subdivided for size or scale. The Wavemaster boat featured on page 158 is an F1 boat. Several competition class courses have been laid down; two examples are shown here.

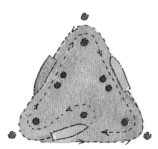

TOP A typical competition racing course

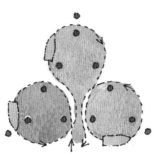

ABOVE A typical course for testing maneuvering skills

BELOW Racing IC-engine powerboats, such as the Jetstream, is an increasingly popular hobby, due to the development of efficient mufflers.

SAFETY RULES FOR POWERBOATS

● Check that powerboat operation is allowed in your chosen pond, lake, or pool.
● Handle boats with care, particularly around the propellers. Check the boat and test the controls and engine on a stand whenever possible.
● Before launching, check that the steering is correctly lined up, with "neutral" (straight ahead) at center on the control stick.
● Warn any bystanders when you are about to start the engine or launch the boat.
● Don't operate in strong currents or in bad weather.
● Don't operate in lakes or ponds with unknown shallows or extensive underwater vegetation.
● Don't operate near people fishing or model yachts.
● Don't send your boat among, or in front of, real boats.
● Don't operate powerboats near wildlife, wildlife habitats, or people swimming or paddling.
● Observe the basic "rules of the road" when operating with other powerboats (see below).
● Check your boat after every operating session. Empty out any water and dry any wet control equipment.
● If there is a lot of water in the boat, check the hull for a puncture and if found, repair it immediately.
● Always recharge on-board ni-cad batteries away from the boat in dry conditions.
● If the battery appears to be running down, head the boat back to the bank. Switching off the motor and letting the boat run under its own momentum can help to conserve power.

RULES OF THE ROAD FOR POWERBOATS

● When two boats are approaching each other head on, each keeps to starboard (right).
● When proceeding up narrow waterways with boats moving in both directions, keep to starboard.
● When two boats are approaching at an angle, or are on courses to cross each other's path at an angle, the boat which has the other to starboard gives way.
● When one boat is overtaking another, the boat overtaking must keep clear.
● A powerboat gives way to a sailing boat irrespective of the above rules.
● If in doubt when several boats are converging on each other, fall back and stand off or run at low speed until the course is clear. This is also relevant if you are operating from a far bank and do not have a clear view of all the boats in the area.

Easy and quick to assemble, the Robbe Laser powerboat is an excellent choice for beginners interested in high-speed models.

BEGINNER'S POWERBOAT

- ### *Watertight hull*
- ### *Powerful performance*
- ### *Matching control set available*

Radio-controlled powerboats are less complex than airplanes or cars in their construction and control equipment. However, you may hesitate to build a model boat for a variety of reasons. You may be afraid of sinking the boat and losing control of it completely. You may be put off by the size of these impressive powerboats. Or you may have tried powerboats but experienced many problems, such as how to keep the boat stable with the control equipment in the hull, how to prevent the boat from filling up with water and sinking, and how to keep the control equipment from getting wet and corroded. Recently, however, kit manufacturers have started to attend to these problems by completing many of the key construction stages for you. The Robbe Laser powerboat is a good example of this type of innovative model that is ideal for the beginner.

The Laser kit

The Laser comes in conventional kit form, but there is less to do than usual because the key stages are completed already. The hull and decks are factory-joined and trimmed, giving a watertight hull. (Access to the motor and control equipment is through a deck opening in the cockpit area.) The kit also comes with a big 500 series electric motor ready-installed and wired. The stern tube, propeller shaft (another area that must be watertight), and the motor controls (forward/reverse) are also in place and wired up. You will need to add the control equipment, but the location ▶

TOOLS YOU WILL NEED

Small screwdrivers
Small scissors
Emery board

1 *The hull comes with the electric motor, shaft, stern tube, and propeller ready-installed and wired up. Fit the rudder servo into position using double-sided adhesive tape, as shown here.*

2 Connect the battery pack supplied in the control set to the receiver, which is already in place inside the hull. All the connections are clip-fits which fit only one way, so no wrong connections are possible. Be sure to keep your fingers clear of the stern, because the propeller may start if the battery has a charge in it. If this happens, stop the motor turning the screw head in the top of the speed controller until the propeller stops.

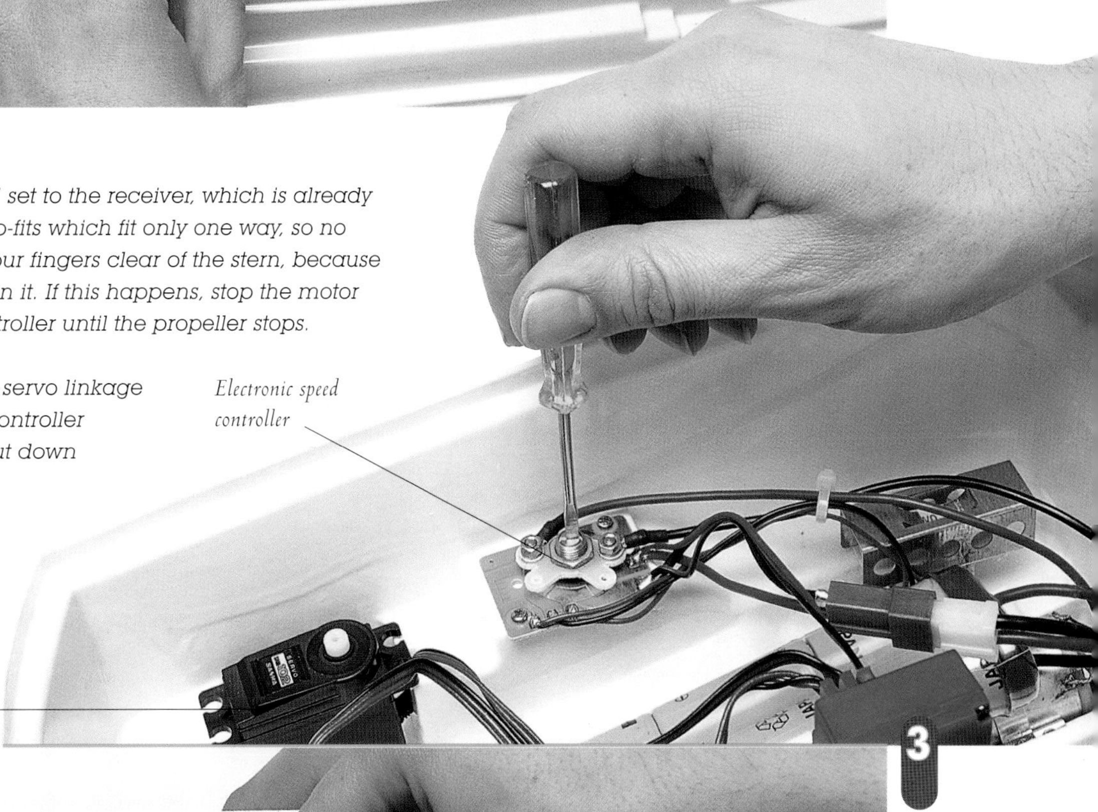

3 Fit the electronic speed controller and screw the servo linkage to it. Equivalent to a throttle, the electronic speed controller controls the running speed of the electric motor. Cut down the star-shaped servo horns to a single arm.

Electronic speed controller

4 Connect the steering servo linkage to the rudder tiller, making sure that the rudder is centered in the center position of the servo. If it is not centered, there will be an incorrect bias in the steering.

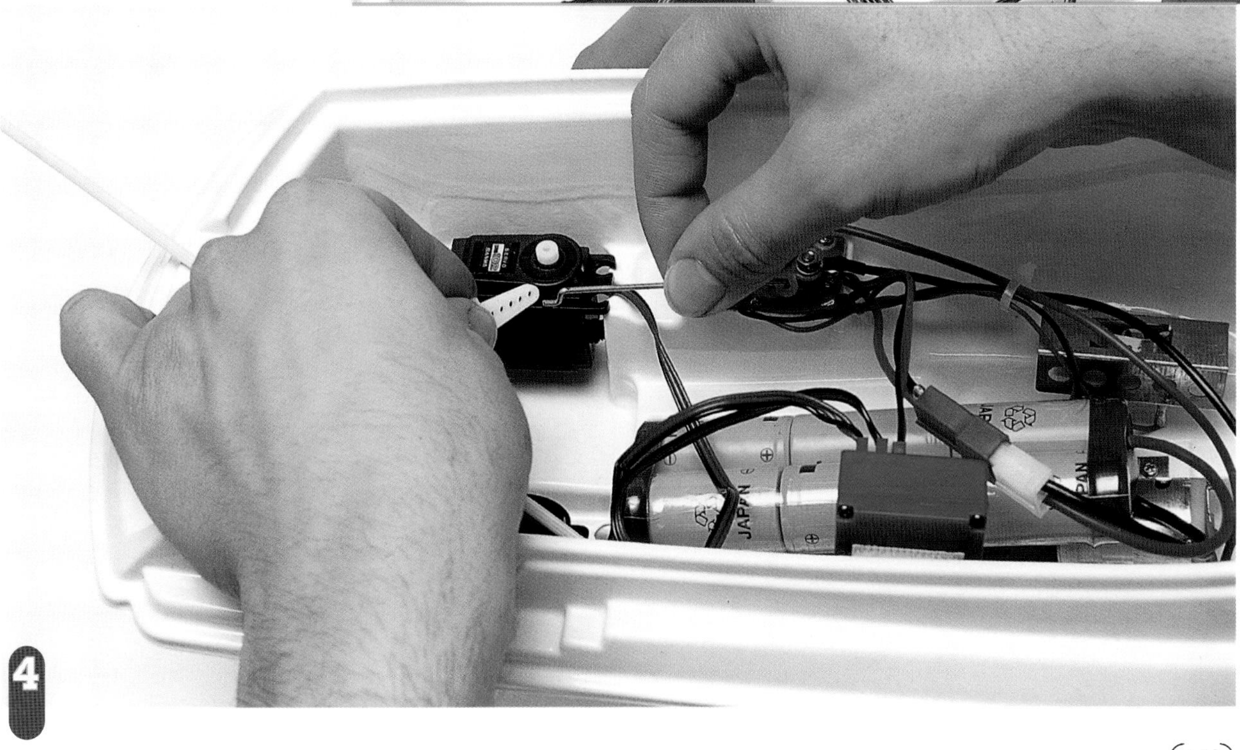

Servo

and installation work is simple and straightforward. The Laser makes up into a lean, keen model, around 2ft 6in (75cm) long, with a good performance. Another boat in the Robbe range with the same specifications as the Laser but with a different hull form (an off-shore racer) is the Razor. The Razor involves the same work as the Laser with the added bonus that the control equipment is also factory-installed.

The controls

The Laser uses simple two-channel radio control with one servo working the rudder for steering and the other controlling the engine speed. A complete control set, known as Control Set 500E, is produced specifically for the boat. This set includes everything you need to complete the model and get it running, making it easy for a beginner.

Test the control set up by holding the boat, switching on the power supply, and checking both steering control and speed while you hold the boat horizontal. Keep your fingers clear of prop and rudder. It may be easier to ask a friend to work the transmitter while you hold the boat.

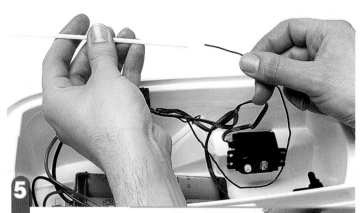

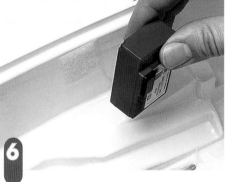

5 Thread the aerial wire through its hollow holder. The aerial is already attached to the receiver, and the holder keeps it up in the air when assembly is complete.

6 Use the double-sided adhesive tape supplied in the kit to secure the receiver against the side of the cockpit well. In this way you can easily access the control components for maintenance or replacement.

7 Stick strips of Velcro to the floor of the well to hold the battery pack in place. In this way the battery pack can be easily removed for recharging before every operating session. A battery charger is included in the Control Set. The complete control installation in the well of the boat is shown here. Note the servo linkage to the rudder and the aerial holder. The on/off switch for the power supply is lower left.

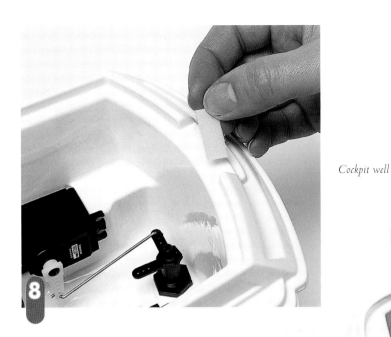

Cockpit well

Cockpit top

8 The control equipment is kept watertight by the cosmetic cockpit top. Use the Velcro provided with the kit to hold the top in place—you will need to be able to remove it easily to get access to the control gear. Note the servo controlling the rudder and its linkage to the rudder head/tiller.

9 When all the control gear is in place, check once more that all the connections are correct, then install the batteries in the transmitter, switch on, and check that the motor and all the controls work. Position the cosmetic cockpit top and press into place on the Velcro strips placed in step 8.

10 Plug the wrap-around windshield molding included in the kit into place.

11 Clip the pre-molded decorative spreader included in the kit into place behind the cockpit.

12 The powerboat kit comes with a sheet of self-adhesive decals, including a cockpit control panel, dummy access door to the engine compartment, non-skid floor panels, side flashes, and names.

Using small scissors, trim each piece and then carefully position on the boat, following the diagrams in the instruction booklet provided with the kit. Rub the decals down with a damp, slightly soapy cloth, pushing from the center toward the edges to eliminate air bubbles.

Control panel decal

Floor panel decal

Running the Laser at speed on a lake is a thrilling experience. As many of the difficult construction stages crucial to the success of a boat model are done for you in the factory, a successful working model is almost guaranteed.

THE FINISHED MODEL
The completed Laser with all its decals in place is an impressive model. Note the dummy side vents for the front engine compartment.

SEE ALSO
Control equipment for
  powerboats, p.150
Introduction to powerboats,
  pp.146–151
Powerboat racing, p.151
Rules of the "road" for
  powerboats, p.151
Safety rules for powerboats, p.151

The Wavemaster by Kyosho is a scale model of an F-1 racing boat. It has a dramatic performance just like the real racing boat but is straightforward and interesting to make.

AN F1 RACING BOAT

• *Easy to assemble*

• *Two-channel radio control*

This high-performance racing boat, the Wavemaster, is a scale model of a typical F1 racing boat frequently seen in modern competitions. With its tunnel hull, it is exciting to look at and equally exciting to operate on a pond or boating lake. The hull traps a cushion of air between the boat and the surface of the water, thus reducing resistance and enabling extra speed. Operating at high speed is particularly exciting as the boat "lifts" its hull and digs in the propeller.

Despite its complex design, the boat is surprisingly straightforward to assemble and is suitable for a beginner. The hull comes in the form of an ABS vacuum-form molding which assembles easily and is very tough. However, what really makes the model so simple to make is the Dolphin 550 outboard motor that comes preassembled and ready to install. The Wavemaster requires two-channel radio control and two servos, which are easy to install. A 7.2V ni-cad battery provides the power. You will need to purchase the batteries and radio-control equipment separately but these will be readily available from your local hobby store. The striking decorations are included with the kit in the form of self-adhesive decals.

TOOLS AND EQUIPMENT
Large and small cross-headed screwdrivers
Pliers
Emery board or sandpaper (to clean any rough edges)

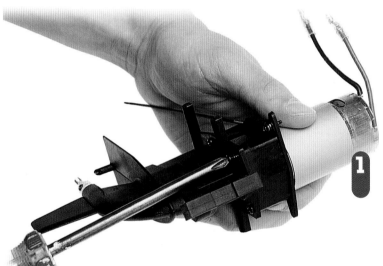

1 *After checking all the kit parts, assemble the motor and propeller first. Simply screw the Dolphin 550 electric motor to the top of the dummy outboard motor housing, which is a neat, hard plastic molding.*

2 *The lower hull containing the sponsons is a single molding and you may need to clean up its edges. Affix the self-adhesive decals to the top of the sponsons. Then screw the mounting plate for the outboard motor at the rear of the hull.*

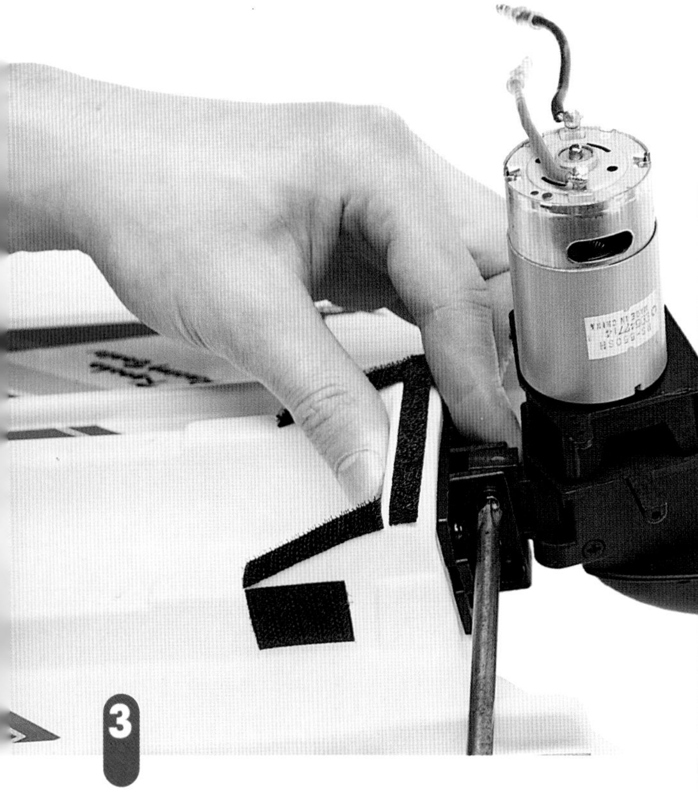

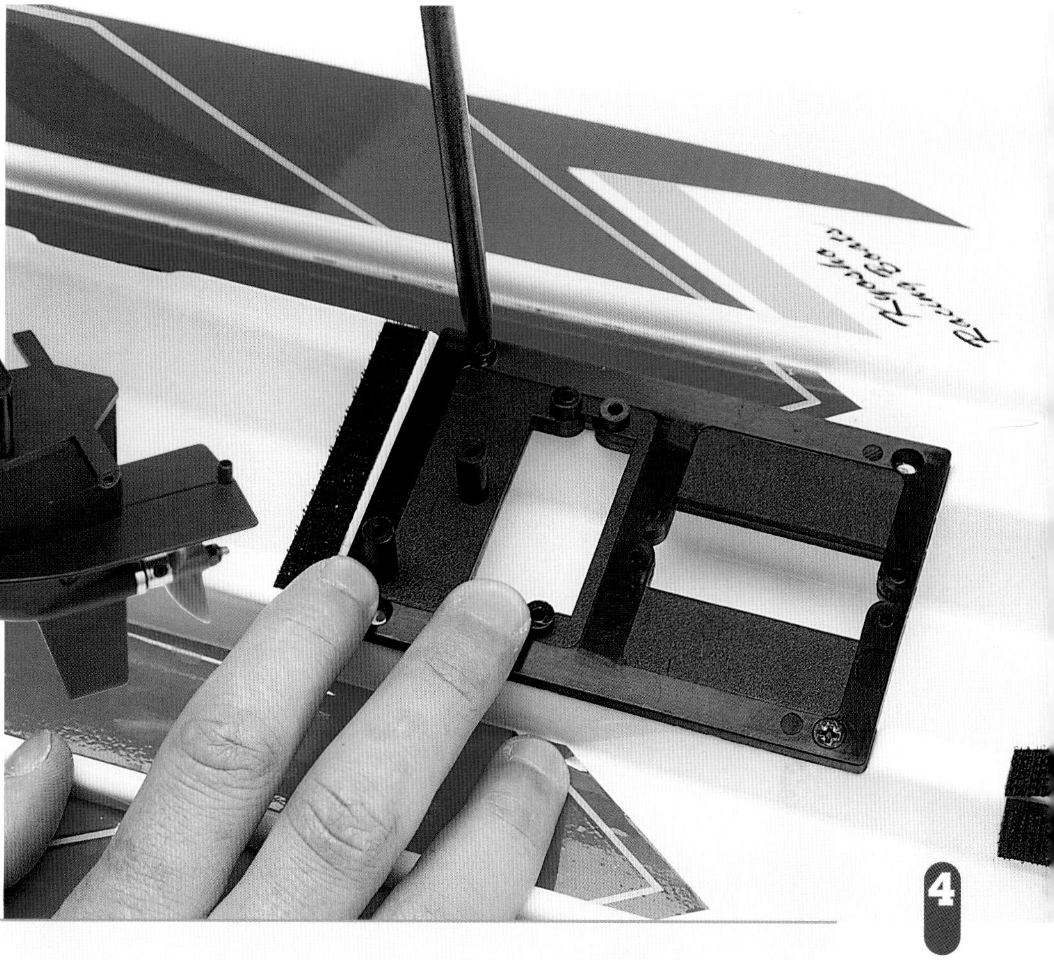

3 Now that the mounting plate is firmly screwed on at the stern, put the propeller in place and assemble the motor. Check that the motor runs freely. The motor assembly should be quick and trouble-free, since it is extremely well designed.

4 Locate the plastic "tray" on top of the lower hull which will hold the control equipment. Screw it into place using the hole positions marked on the molded plastic lower hull.

5 Locate the steering servo and attach the wire rod linkage to actuate the steering to the outboard motor—this pivots as a complete unit. Attach the inboard end to the yoke on the servo. Before screwing the servo into place, try the arrangement for fit, as shown here.

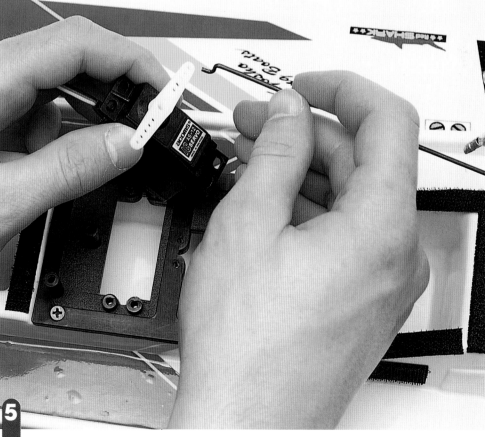

6 Screw the steering servo into position on the control equipment tray with the steering linkage hooked into the yoke on the servo. It is essential that the yoke, the outboard motor, and the linkage rod are all centered when this is done, so that steering is accurate when using radio control.

7 Now screw the other control servo into place, the opposite way around from the steering servo. This servo controls the on/off switch for the motor, so that the boat can be stopped and started offshore. It will be controlled by the second channel.

8 Screw the on/off switch for the motor into place forward of the two servos. Cut the yoke for the second servo to length, as shown, then add the connecting link to the switch.

9 Wire the on/off switch up to the motor, then connect the switch and servos to the receiver, as shown here. Hold the receiver with one hand while you make the connections.

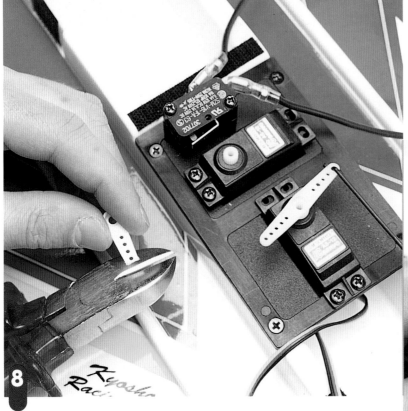

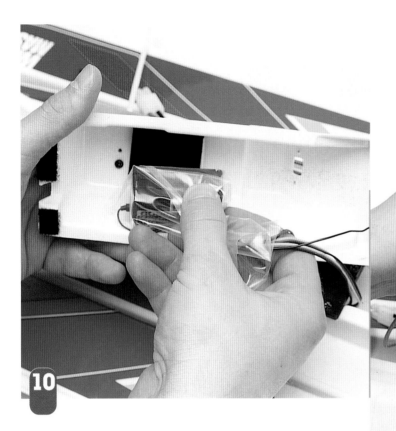

10 *It is essential that the receiver on a powerboat is completely watertight. You can ensure this by inserting the receiver into a clear plastic bag and fastening the bag with elastic bands or adhesive tape. The receiver fits under the control equipment tray.*

11 *The final control item is the ni-cad battery pack. Connect it up and fit it into the remaining recess in the control equipment tray. Check all the connections and then hold up the model and test the control functions using the transmitter. If all is well, you can place the top cockpit/fairing molding over the lower hull to conceal the control equipment.*

12 *Add the top housing to the rear outboard motor assembly to cover the electric motor and keep it watertight. You can now add the extensive self-adhesive decals included in the kit to the superstructure. Affix the cockpit windows also.*

The completed Wavemaster is a sleek-looking modern competition F1 powerboat. With its impressive size— over 2ft (60cm) long—it performs like the real thing and is spectacular when operated at high speed.

PROLONGING SAILING LIFE

This model has a well-sealed outboard motor and a watertight hull. However, you should always clean and dry your model and control equipment carefully after each session.

SEE ALSO

Introduction to powerboats, pp.146–151
Racing and safety tips, p.151

Model yachts are undoubtedly the gentlest and most graceful of all models that can be controlled by radio. The sight of a miniature yacht sailing across a lake or pond, powered by one of nature's own power sources, the wind, never fails to fascinate. The yacht's progress is quite slow and leisurely, except in a strong wind, and you may be tempted to take up the hobby after watching modelers enjoying a sailing session at the lakeside.

INTRODUCTION TO RADIO-CONTROLLED YACHTS

In the past all model yachts were free sailing, but, gradually, the desire to hold yacht races led to the introduction of new features allowing greater control of the yacht. An early development was automatic steering, controlled by a wind vane and pre-set before the boat was launched into the water, which, with luck, kept the model (and its competitors) on the same course to the far side of the pond. Radio control was obviously a big advance on the limitations of wind-vane steering, allowing modelers to control their boats from the water's edge. Progress was slow until the 1960s, as radio equipment was too heavy and bulky to fit into even a big model. Over the years technological progress has transformed radio-control equipment, reducing it to small lightweight "black boxes" that are perfect for model yachts.

CHOOSING A YACHT KIT

Many radio-controlled model yacht kits are available, ranging from simple yachts to immensely complicated and sophisticated racing yachts. As with most other radio-controlled models, it is better to start off with a straightforward yacht kit, gain sailing experience with that, then try a more complex kit and progress from there. Note that very few of the mass-produced, easily built yachts conform to the established model yacht racing classes (see page 174). In many cases, the modern kits are models of actual boats, with highly detailed features, such as miniature crew figures.

ABOVE Running before the wind in classic manner is the Kyosho Seawind, a model of the Americas Cup type. Note how the sails are almost at right angles to the yacht.

RIGHT The schooner-rigged Atlantis, manufactured by Robbe, is a magnificent model, and ideal for the more experienced enthusiast.

BELOW The Kyosho Seabreeze with all the control equipment required for a radio-controlled yacht shown in front of it. The complete setup includes transmitter, two servos, battery pack holder, switch and receiver, and the all-important cradle.

Transmitter

Cradle

Battery pack holder

Servos

Switch and receiver

MATERIALS

Model yachts used to be made of traditional material, mostly wood. You can still see some wooden handmade scratch-built models, and there are even a few wood kits on sale from time to time. More recently GRP (fiberglass) has became popular as a material for molded hulls. However, virtually all the popular kits available today have a hull molded in ABS plastic, which is light and durable and lends itself to mass-production methods. Masts and booms are most often of extruded aluminum, and spreaders, mast caps, blocks, and bowsies (tensioners) are in hard plastic. Eyes and hooks are in brass or softer metal; the latter should be handled with care, as they are fragile. Sails in modern kits are usually made of polyester, and they are usually supplied in a finished state, ready to fit. However, in some kits you may simply get a pattern and a roll of fabric from which to make the sails. Note that, with use, original sails may get torn or stretched out of shape and you will need to make new sails. Racing enthusiasts often have several different types of sails to suit different wind conditions. Sufficient cord for rigging your model is normally supplied in the kit.

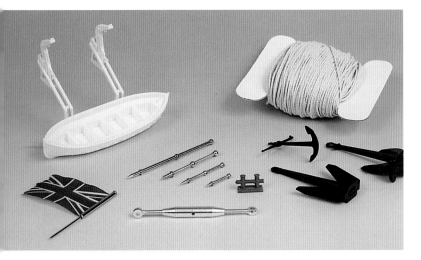

ABOVE Numerous accessories for model yacht detailing, repair, and maintenance are available from specialist hobby stores.

Foresail

Mainsail

88

Comtesse

Keel

NAUTICAL AND SAILING TERMS

Once you start operating on the water, you will encounter a number of nautical terms.
A knowledge of these terms is also useful for understanding kit instructions and model-yachting magazines.

port	left
starboard	right
port tack	sailing with the wind on the port side
starboard tack	sailing with the wind on the starboard side
fore	toward the front of the boat
aft	toward the back of the boat
bow	the front of the boat; also port bow and starboard bow
stern	the rear of the boat
midships	the centerline of the boat
gunwale	the edge between the deck and the hull side
draft	the depth of the hull below the waterline
freeboard	the height of the hull above the waterline
standing rigging	the rigging that is fixed in place as a support for the mast, etc.
running rigging	the rigging that moves (can be hauled) to move the booms, etc.
stay	standing rigging to support the mast; cf. forestay and backstay
shrouds	standing rigging port and starboard to support the mast—may be adjustable for tension
halyards	running rigging for hoisting the sails, etc.
sheets	running rigging for controlling the sails
outhaul	running rigging for pulling the sail out to the fullest extent on the boom
boom	pole to which the foot of the sail is laced or attached
quarters	rear corners of a boat; also port quarter and starboard quarter
bowsie	tensioner (model yachts only)
downhaul	running rigging to pull the sail downward to the fullest tension
kicking strap (vang)	length of rigging from boom to foot of mast to prevent the sail "bellying" and the boom lifting in high winds
tack	forward corner of a sail
clew	rear corner of a sail
luff	leading edge of a sail
leech	rear edge of a sail
head	top of a sail
foot	bottom of a sail
mainsail	the largest sail in a typical yacht rig
foresail (jib sail)	the smaller forward sail in a typical yacht rig
flying jib	second smaller foresail, sometimes carried above the jib sail
Bermuda rig	the most common sail arrangement found on a model yacht, though on real yachts there are many more options
Genoa jib	foresail of larger area, sometimes rigged in place of the jib on some classes
keel	the vertical extension, usually ballasted, below a yacht's hull to give it stability and purchase against the wind
centerboard	retractable keel common on real small yachts, occasionally found on models such as sailing catamarans or trimarans
rudder	the moving attachment below water used to steer the boat
tiller	the arm at the top of the rudder used to control its direction
on the beam	to the side; e.g. port beam and starboard beam

BELOW There are many small but attractive cruising yachts, such as the Robbe Contesse, that are ideal for newcomers to radio-controlled model yachts.

Rudder

ABOVE Before launching a yacht, check that all rigging is correctly tensioned and unobstructed. Check also that the rudder and sail servos repond to commands from the transmitter.

HOW A YACHT SAILS

A model yacht sails just like a real yacht, and with radio control you can control it from the waterside just as if you were aboard. The sails on a yacht have a similar function to the wings on an airplane. The curve in the sail gives an airfoil type of section. Air pressure on the windward side pushes the sail (and hence the boat) along against the lower air pressure on the other side of the sail. Because the wind must come on the sail at an angle to get this effect, a yacht cannot sail directly into the wind. If it does, wind spills from the sails and they flap uselessly. In practice most yachts cannot sail closer than 45° to the wind. This position, when the sails are close into the line of the boat, is known as close-hauled.

As the yacht changes its heading from the wind, the wind comes in on the beam. This position, with the sails let out so that they are at an angle to the boat, is known as reaching. Coming farther off the wind, so that it is coming from behind the beam, the sails are almost at a right-angle to the yacht, in a position known as running. There is always a tendency to sail a boat as close to the wind as possible when tacking, and if you come too close to the wind, the fore edge

of the sail, and then the entire sail, will start to flutter. A good sailor will steer the boat farther off the wind as soon as any flutter is seen. A yacht sails most efficiently when the sails are full and trimmed to the optimum angle to take advantage of the wind direction relative to the course. Radio control allows you to trim the sails constantly to achieve this. To sail your model competently, you will need to master two essential maneuvers: tacking and jibing.

Tacking

Since a yacht cannot sail directly into wind, it must tack to progress forward into the wind. This means it sails one leg to port at 45° to the wind, then changes direction by turning its bow through the wind, so that the wind fills the sails on the opposite side, and sails another leg to starboard. Then it changes direction again, making forward progress into the wind by a zigzag course. Each tack should be as long as possible to save time. Speed falls off as the bow turns through the wind, and an experienced sailor will come off the wind for a second or two to increase speed before turning the bow smartly though the wind. If this is done correctly, the wind will fill the sails on the other side and the yacht will go off on the opposite tack. If it is done too slowly, you may lose the wind and stay head to wind with the sails flapping until the bow falls away and you pick up the wind again. A yacht stuck like this is said to be "in irons."

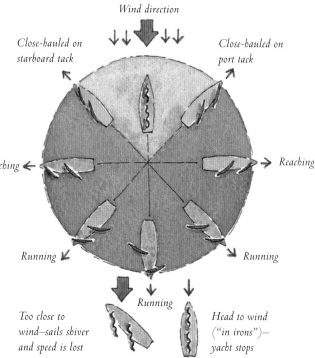

Sail and course positions relative to wind direction

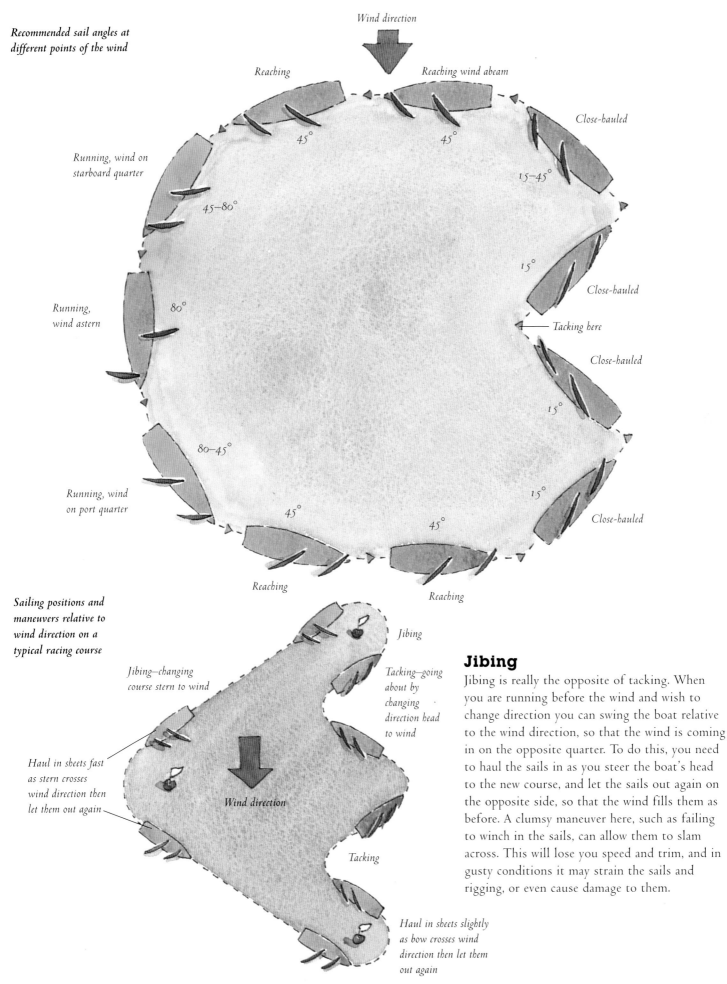

Recommended sail angles at different points of the wind

Wind direction

Reaching

Reaching wind abeam

Close-hauled

45°

45°

Running, wind on starboard quarter

15–45°

45–80°

15°

Close-hauled

Running, wind astern

80°

Tacking here

Close-hauled

15°

80–45°

15°

Running, wind on port quarter

45°

15°

Close-hauled

Reaching

45°

Reaching

Sailing positions and maneuvers relative to wind direction on a typical racing course

Jibing

Jibing—changing course stern to wind

Tacking—going about by changing direction head to wind

Haul in sheets fast as stern crosses wind direction then let them out again

Wind direction

Tacking

Haul in sheets slightly as bow crosses wind direction then let them out again

Jibing

Jibing is really the opposite of tacking. When you are running before the wind and wish to change direction you can swing the boat relative to the wind direction, so that the wind is coming in on the opposite quarter. To do this, you need to haul the sails in as you steer the boat's head to the new course, and let the sails out again on the opposite side, so that the wind fills them as before. A clumsy maneuver here, such as failing to winch in the sails, can allow them to slam across. This will lose you speed and trim, and in gusty conditions it may strain the sails and rigging, or even cause damage to them.

THE RADIO CONTROL ELEMENTS

However big and complex a model yacht may be, the radio control elements are similar, although they may vary in degree of sophistication. The two-channel system is used with a standard transmitter. The right stick, moving left and right, controls the rudder, and the left stick, moving up and down, controls the sail winch. Moving the stick up lets the sheets out, while moving it down hauls them in. Some enthusiasts prefer to use a transmitter with a control wheel to control the steering. Some of the simpler yacht kits use single-channel with a control for the rudder only. In this case, the sails are simply set up in a "halfway" position, as is done on a free-sailing yacht. However, some simple yachts have the option of sail control as well, turning them back into two-channel control models.

A receiver unit in the boat takes the signals from the transmitter. This has some sort of aerial, which may simply be a wire led around the cockpit interior or incorporated into the rigging. One good idea sometimes seen on highly tuned racing yachts is to have the backstay doubling as the aerial. Signals via the receiver go to (1) the steering servo which moves the tiller/rudder from side to side, and (2) the sail control unit (sail winch), which hauls the sheets (and thus the sails) in and out to adjust to the wind conditions and the course being sailed. The sail control unit is usually a bar attached to the servo spindle in the smaller, simpler yachts. This bar is set so that

RIGHT This close-up view of the Seawind shows the kicking strap, shrouds, downhaul, and outhaul. The winches on the deck are cosmetic scale dummies.

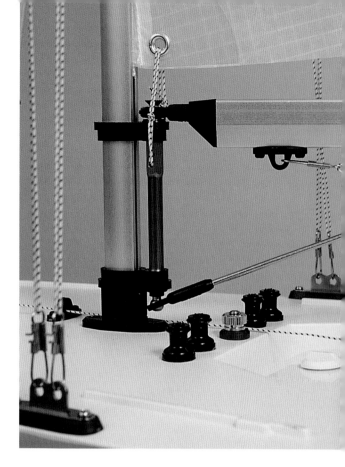

BELOW The most typical layout for yacht radio-control installation is shown here. In top racing yachts, the sail control unit with its long arm, may be replaced by a servo with a rotating drum.

its movements haul in the sheets or let them out. On racing yachts or the bigger kits, the sail control unit may take the form of an actual winch drum on the servo, which rotates to wind the sheets round the drum just like a real winch. This racing servo is a high-torque type, which will require more power to operate, so you will need to have a more powerful battery pack. The fastest working racing servos can haul the sheets

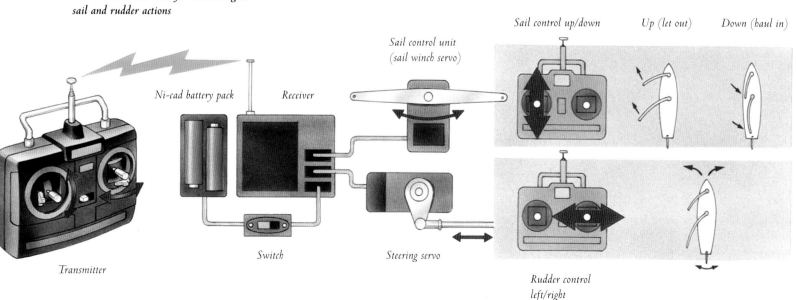

Transmitter movements for controlling the sail and rudder actions

Sail control unit (sail winch servo)

Sail control up/down   Up (let out)   Down (haul in)

Ni-cad battery pack   Receiver

Transmitter

Switch

Steering servo

Rudder control left/right

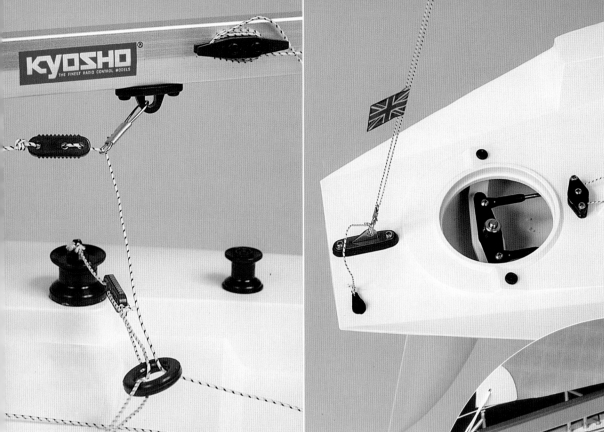

LEFT CENTER *The outhaul and main sheets on the Seawind. The sheet is taken to the sail control unit inside the hull.*

LEFT *A view of the tiller on the Seawind. Note that the watertight access hatch has been removed to allow access for maintenance.*

in and out in as little as three seconds, allowing very quick tacking or jibing in regatta conditions. The smaller yachts with bar-type sail control units use ordinary servos, the same type as used for the steering, so only a basic battery pack is needed. The control system is switched on and off with an accessible power switch.

FITTING THE CONTROL EQUIPMENT

The control equipment normally fits in the cabin/cockpit area. This is covered by the cockpit or part of the cabin roof, which is detachable for easy access. It is important that all this equipment is kept watertight. Good yachts are usually well designed with seals and close fitting slots for the control equipment. However, to protect your equipment more, try covering the battery packs and other elements with transparent plastic bags. Take the wiring out through the ends and then secure the bags with rubber bands. This may lead to condensation, but you can prevent this by removing the access section and any plastic coverings at the end of a sailing session. Then just before you sail again, return it to its "watertight" condition.

As with all radio-controlled models, there is always a risk of loss of control if the equipment, or any part of it, becomes defective during operation. A yacht, however, has the advantage that, even when control is completely lost, it

should sail on like a free-sailing model and eventually reach the far shore, even if this is not where you intended it to go.

LEARNING TO SAIL

The only easy way to gain proficiency in sailing a model yacht is to get out on a pond and practice. Choose a day with a gentle breeze rather than a gusty day. Until you are confident of controlling your model, avoid any area where other yachts are being sailed to prevent accidents or annoyance to other modelers.

ABOVE *The Tamiya Yamaha 40EX, 1:20 scale ocean racer, with its hull cut away to show the control installation inside, including the sail control unit.*

DOS AND DON'TS OF
YACHT SAILING

● Do not sail, or attempt to sail, in strong winds or strong water currents. Avoid river sailing for this reason.

● Sailing in conditions of "no wind" is frustrating, and may also cause damage to the equipment, such as the sails if they are allowed to flap uselessly.

● Do not sail in ponds or lakes with many unknown shallows or extensive vegetable growth. Weeds may catch the keel and in extreme cases could pull a yacht under. Or your model may be stranded out of your reach.

● Check that model sailing boats are not prohibited in your chosen location. Never sail near swimmers, real boats, or wildlife.

● Most model yachts are quite big. Make sure you have enough space to handle your model and that you do not cause a nuisance to any bystander. Always check that nobody is behind you when you handle the model.

● Do not sail near people fishing and never sail in tidal waters where there is any real boat traffic.

● Observe all sailing rules. Avoid colliding with other model yachts either deliberately or accidentally. Collisions can cause expensive damage and much bad feeling.

SAILING PREPARATION AND
MAINTENANCE

Because they are in contact with wind and water, model yachts will need cleaning and checking after each sailing session. Most kits include a cradle in which the boat can rest (and be tilted) for cleaning and tensioning of the rigging, and so on, and setting up prior to sailing. Cradles are also sold by hobby stores specializing in boat models. Specialist suppliers sell rigging cord, blocks, eyes, bowsies, and all the other replacement parts you might need for later repairs. Check the radio-control gear before and after each session, and always recharge the battery pack before a session. If GRP hulls are holed, you can use a resin repair kit available from specialty stores to repair the damage. Modern ABS hulls are very tough, but if they get chipped or dented, try repairing them cosmetically with modeling putty. In the less likely event of an ABS hull being pierced, it is possible to patch it using glue suited to ABS.

ABOVE A cradle is a necessary accessory for all radio-controlled yachts. It can be used to display the model but is also used for maintenance and setting up the yacht on the waterside.

GETTING BETTER PERFORMANCE

If you want to get better performance from your yacht the rigging and sail setting must be perfect. Sagging sails and flopping booms will not help performance. Polishing the hull to a high degree will get an edge on speed, but be sure that any polish you use does not adversely affect the ABS or other plastic that forms the hull. If necessary test the polish first on scrap plastic of the same type from the kit leftovers. To help you judge the wind direction and strength, it is a good idea to tie colored wool or thin ribbon strips to each shroud. Their movements, visible from the shore, will help you to guide your yacht more smoothly.

Racing or identifying number

Mainsail

Foresail

3215

Rudder

RIGHT The Kyosho Fairwind with its ABS-molded hull is a typical example of a modern model yacht built from a kit. All the parts, including the sails, come ready for assembly.

Cradle

FAIRWIND II

KYOSHO

FAIRWIND II

YACHT-RACING CLASSES

At a model-yacht regatta you are likely to see many classic model yachts from the racing classes listed below. Most classes predate radio control—classes including radio-controlled yachts are indicated by the prefix R. The rules concerning dimensions, sail area, displacement, etc. apply whether the yacht is free sailing or radio-controlled.

R575

This is an individual class (but not officially recognized) for radio-controlled yachts. The class originated from the decision of several small British kit firms to conform to a small size formula for a "one design" racing yacht with a length of 22.5in (575mm).

R36R

A famous British racing class dating from 1930. The models in this class are restricted to 36in (915mm) length, 11in (280mm) hull depth, and 9in (230mm) beam. Maximum weight 12lbs (5.5kg), unlimited sail area.

R1M (or Meter)

This is a more modern international equivalent to the R36R. The hull length is $39^{1}/_{2}$in (1m), and there is a maximum and minimum weight limit and a choice of three different sail rigs.

RC-Marblehead

This classic model yacht class was established in 1932 and designed by members of the Marblehead Model Yacht Club of Massachusetts. The length is 50in (1.27m) maximum, and sail area must not exceed 800in^2 (0.516m^2). There is no limit on beam, freeboard, displacement, or sail plan, which has led to the appearance of imaginative, competitive designs.

R10 (10 Rater)

The rating is determined by a simple formula of the waterline length (in meters) multiplied by sail area (in square meters) multiplied by 8 to equal 10. As the length can vary against the sail area, modelers have come up with a wide variety of designs in an attempt to achieve optimum sailing performance.

R6M (6 Meter)

This 6-rater type is similar to the R10 above except the waterline length multiplied by sail area multiplied by 8 must equal 6. It is largely used for free-sailing models, although there are some radio-controlled versions. The hull of the R6M must be 55–60in (1.40–1.52m) long.

EC R12M

This is an American design (EC=East Coast) mostly found in the United States. The hull length must be 58–60in (1.47–1.52m), and the displacement must not exceed 26.5lbs (12kg). This is a very handsome type and is used for the Mini-Americas Cup races.

RA (International A)

This is the top class, designed in 1923, which features the biggest and grandest of all model yachts. There are strict rules concerning displacement, sail area, and other aspects. The hull must be 65–90in (1.65–2.3m) long, and the weight must not exceed 55lbs (25kg).

LEFT The Tamiya Yamaha 40EX 1:20 scale ocean racer is 24in (600mm) long and stands 47in (1.11m) in overall height. The mast and booms are of aluminum and the sails of lightweight spincloth, as used on the actual yacht.

STARTING YACHT RACING

If you want to get involved in serious yacht racing it is best to join a club. Look in the various model-boat and yacht magazines for more information about clubs and to find out what yachts they use—they may race on a class basis (see p.174) or they may organize races for a particular model. In addition, leading kit makers, such as Robbe and Kyosho, sometimes hold regattas for the "one design" model yachts they produce in kit form. You can, of course, get together with a couple of friends with models made from the same kit and hold an informal race. This can be great fun, although it won't compare with properly organized class racing.

Official racing

Official model-yacht competitions are held under the auspices of a long-established international body, the International Model Yacht Racing Union. One of the most famous events organized by the IMYRU in recent years was the Mini-Americas Cup series for model yachts that ran concurrently with the Americas Cup events. The IMYRU also draws up the rules for international racing classes, each of which has its own set of rules. In addition, national organizations such as the American Model Yachting Association and the Model Yachting Association in Great Britain run competitions and events in their own countries and draw up rules for classes that may be recognized nationally but not internationally. Before yachts can take part in national or international competitions, they must be inspected and issued with a certificate proving that they conform to class rules.

Model-yacht racing rules

Both the national model-yacht organizations and the IMYRU have detailed rules for races held under their auspices; copies of the rules can be obtained from your national organization. The rules are very similar to the rules for real yacht racing. It is important to memorize the key points and observe them, even if you are only racing informally with a friend or merely sailing randomly for fun when there are other model yachts present on the same pond. The key points are as follows:

● Yachts on port tack give way to yachts on starboard tack.
● Upwind yachts keep clear of downwind yachts.
● Tacking or jibing yachts keep clear of yachts on a tacking course.
● When rounding course marks (buoys), outside yachts must leave room for inside yachts to clear the mark also.
● Yachts that cross the start line before the start signal must circle round and cross the start line again.
● Yachts that touch a course mark (buoy) while rounding it must sail round it twice (clear of any other competing yachts) before proceeding on the course.
● Yachts that infringe any of the rules will be disqualified.

RIGHT A good example of a model R1M (Meter) class competition yacht, the Robbe Windstar has a streamlined ABS hull and is 39¹/₂in (1m) long. Three sets of sails: heavy weather (shown here), standard, and light are available.

Many yachts are difficult to build, and require care in assembly. However, the Robbe Dolphin is one of the easiest yachts to make and operate.

A GRACEFUL YACHT

- *Easy to build*
- *Ready-made sails*
- *Matching control set available*
- *Yacht cradle included*

A wide variety of radio-controlled yacht kits are available, although some are complex and expensive, which might deter beginners with the urge to sail. Some kits require a considerable amount of work—sometimes you may have to make the sails yourself—and are long-term projects. If you are looking for a winter-long project, for example, time-consuming yacht kits are ideal. On the other hand, if you are not too confident of your skills, or are keen to get a yacht model afloat and under sail as quickly as possible, then the simple Robbe Dolphin yacht is an ideal choice.

Intended for beginners, the Robbe Dolphin is so simple that the manufacturers optimistically claim that you can "start it today, sail it tomorrow." Although this would only be true if you spent an uninterrupted 24 hours assembling the model, the model can be finished in a few days following the instructions supplied with the kit.

TOOLS YOU WILL NEED

Hand drill (supplied in the kit)
Mini-drill
Small screwdriver (for crosshead screws)
Large screwdriver (for crosshead screws)
Flatnose pliers
Craft knife
Strong thread

Building the yacht cradle

The first task, not covered in the illustrated sequence here, is the assembly of the yacht cradle supplied with the kit. The cradle makes a perfect stand, not only for the display of the finished model, but also for holding the hull while you are working the model. Later, the cradle will be useful for holding the completed model at the waterside, while you are preparing to sail. The ▶

1 *Start by fitting the hooks that hold the mast stays. Glue the strengthening piece (supplied with the kit) inside the hull on each side. Pierce the holes for the hooks in the marked positions, using the hand drill provided in the kit. The hooks are a tight fit, and you may need to use pliers to screw them in. However, do this very gently, since the metal is soft and can easily be twisted or broken.*

2 *Fill the keel with the ballast weight (shot) supplied in the kit. This helps to prevent the model from capsizing—a feature useful for beginners. The ballast is held in place by a plastic "plug"— glue it into the keel recess with a good coating of glue to ensure it is watertight. Glue the mast step to the keel plug, and then glue the mast housing—a brass rod supplied in the kit—firmly into the mast step through the cabin roof, as shown here.*

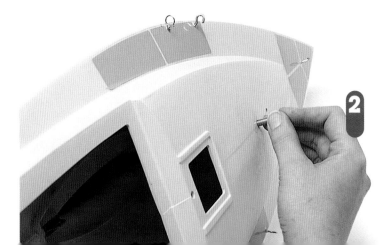

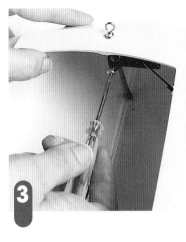

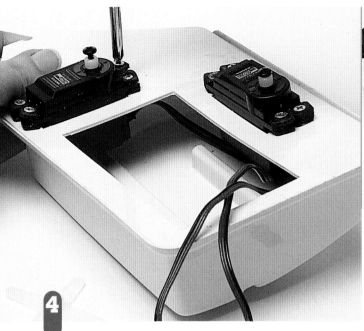

3 *Glue the rudder post inside the stern, and insert the rudder from below. It is held in place by the tiller yoke—screw it into place at right angles to the rudder, as shown here. It is essential to achieve the correct angle.*

4 Screw the two servos supplied in the control set into the assembled tray that holds all the control gear. Hold the servos tight with the rubber grommets from the kit, and, using the mini-drill, drill holes in the marked positions to hold the servos in place. You will need to find small screws and a larger screwdriver to secure the servos.

5 *Now that the servos are in place, wire the receiver, also supplied in the control set, to the servos. Use the double-sided tape supplied in the kit to hold the receiver in place on the bottom of the tray.*

6 *Check the battery pack supplied in the control set. Then wire it to the receiver, and put it into the remaining space in the tray. Use sponge rubber to hold it in the tray, but don't fix it down, since the battery pack needs to be removed for recharging. The aerial attached to the receiver is in the form of a wire—this will be taped inside the cockpit tray later. The control tray is now complete with all the components installed. Later, the tiller will be attached to the left servo, and the sail winch to the upper servo.*

7 *Glue the control tray securely into the cockpit opening, close up to the rudder post. When the glue is set, screw the tiller arm directly into the spindle of the servo, checking that the rudder is centered at the middle position of the servo. This servo moves the rudder from side to side.*

8 *Supplied in the kit is a plastic arm, to which the sail sheets will be attached later. Attach this arm to the second (sail) servo. This will haul the sheets in and let them out under radio command. Drill holes in the marked positions, using the mini-drill, and then screw the arm to the servo spindle at the correct angle.*

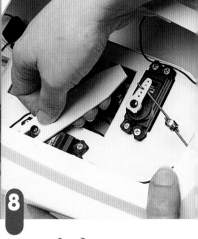

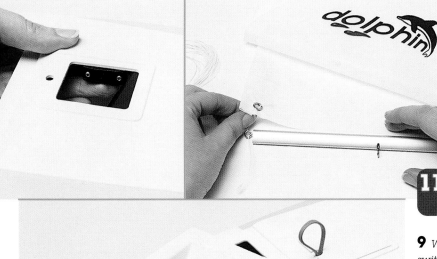

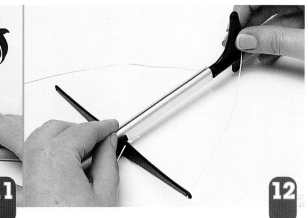

9 Wire the receiver and battery pack to the battery on/off switch, the last part of the control system. Use glue or double-sided adhesive tape to affix the switch unit under the cabin roof by the hatchway. Later, a hatch will be taped into position to keep the interior watertight. The hatch can be opened to allow access to the switch.

10 Add the decorative finishing touches to the yacht, using the self-adhesive decals supplied with the kit. Stick seat cushions on the cockpit well, and add the yacht's name, cabin windows, and non-skid areas on the cockpit floor and deck. When the cockpit well is fully decorated, fit it into the cockpit area to cover the control tray, and keep the boat watertight. If it doesn't fit tightly, use double-sided tape to hold it securely in place.

11 Although this is a straightforward model, the rigging of all yacht models is a complex and time-consuming task. Fortunately, the Dolphin comes with ready-made sails. Here, the main sail is being laced to the boom. Secure the elastic band visible in the cab roof to a ring glued above the keel. It will be used later as part of the kicking strap arrangement to prevent the sail and boom lifting too high, if struck by a squall.

12 Now rig the main mast. Attach the masthead and the spreader moldings to the tall main mast, and thread the rigging cord provided in the kit in place to form the stays. These will be secured to the hooks that were screwed into the hull at the first stage of construction. Though simpler than the rigging on many yachts, the Dolphin's rigging plan is still quite complex, and should not to be rushed. The kit instructions are extremely detailed, dividing the task into standing rigging (the stays and shrouds that support the mast) and the running rigging (the sheets that are taken to the arm of the sailwinch servo). Study the diagrams carefully, since incorrect rigging will make your yacht perform poorly and may also damage the mast and boom.

cradle consists of two plastic fore and aft supports with wood sides which join the two. Glue the cradle together with strong contact glue and allow to set—you can use clothes pegs to hold the parts together while the glue is setting. Cover the supports with foam rubber strips, so that the cradle won't scratch the hull.

Operating the Dolphin

The Dolphin is a basic, two-channel radio-controlled model. One servo operates the steering, and another controls the sail windlass. Although the control system is quite conventional, some "trimming" will be necessary. Begin by getting the tiller set up, so that the rudder is in the "straight-ahead" position, when the stick on the transmitter is centered. Then make sure that you have the correct length of sheet attached to the windlass arm, so that the sails are fully out when running.

All the control equipment necessary for operating the Dolphin is available in a conveniently boxed "Start Set," which must be purchased separately from the kit. Included in the set are a battery pack for the yacht, a battery charger, receiver, and two servos. The only other extra item of equipment that you need to buy before you set sail, is eight AA size batteries for the transmitter.

LEFT: *The completed model is graceful and impressive. The mast is over 3ft (90cm) high, but it fits over the housing and is held in place by the stays, so the model can be transported without its mast. You can then fit it at the lakeside as you prepare the yacht for launching.*

SAILING THE YACHT
Choose a calm day for your first outing, so that you can get a feel for the sailing characteristics of your model. Always remember to check the rigging thoroughly at the waterside before setting sail, to ensure a smooth performance.

SEE ALSO
Tacking and jibing, pp.168–169
Dos and don'ts of yacht sailing,
  p.172
Introduction to model yachts,
  pp.164–175
Nautical and sailing terms, p.167
Sailing and racing rules, p.175

Boats come in a surprising variety of sizes, types, and styles, ranging from mini powerboats to elegant yachts with full rigging. Also included in this section are model submarines, which can be operated underwater, offering perhaps the ultimate challenge to radio-control enthusiasts. Examples of exciting model craft of all types, some suitable for beginners, others for the experienced, are shown here.

GALLERY

ABOVE *An unusual model, the Robbe Sea Jet is a large-scale replica of a jet ski. The model is 24in (590mm) long, and the hull comes pre-colored, sealed, and trimmed. All the power system components are included in the kit, and assembly takes only a few minutes. Because it is well sealed it will not sink, and the model has a lively performance, jumping from waves and even doing Eskimo rolls.*

LEFT *About 1:25 scale, the Robbe harbor tug Odin looks extremely realistic, especially on smooth water with no "out of scale" waves. In spite of its sophisticated appearance, the model uses only a standard electric motor and simple two-channel radio control.*

ABOVE *Not all radio-controlled sailing models are yachts. This handsome scratch-built boat is a model of a traditional River Thames sailing barge that was once a common sight carrying cargo in the Thames estuary and around the east coast of England.*

LEFT *The Seawind by Kyosho is modeled on the Americas Cup yachts. It has a tough ABS hull and aluminum mast and booms. The model is highly detailed with scale deck fittings, including a dummy steering wheel. The yacht is easy to maintain, since the internal control tray is a modular unit. It is also easy to transport, because the mast, keel, and rudder can all be detached. The handsome stand is included with the kit.*

ABOVE *This classic model racing-boat is the Kyosho Jet Stream GP. Reaching up to 22mph (35kph), it is powered by a recoil start water-cooled IC marine engine. The model has a strong, smooth ABS plastic hull, and the motor and control equipment are all carried on a platform below the dummy cockpit and engine cover, which detaches for interior access. The three crew members are also included in the kit. An electric motor version, the Jet Stream EP, is also available*

LEFT *The Graupner Mini Sprint is a high-performance powerboat that is simple to build and operate. Although it is only 16in (400mm) long, it runs extremely well owing to its Speed 400 electric motor and aerodynamic hull shape. It is made for two-channel radio control but only has one servo for steering, and a soft switch for starting and stopping the motor.*

BELOW *This Precedent model of the
Malaysian Navy gas turbine patrol boat
Perkasa is available in two different scales:
1:32 and 1:24. There is also a choice of a
wood or fiberglass hull. This boat is ideal for
advanced modelers, because you must choose
the form of propulsion and control yourself
—there are no recommendations from the
manufacturers. However, there is plenty of
space inside for the hull for the necessary
control equipment and either an electric
motor or a water-cooled IC engine.*

ABOVE *The Harbor Star is an elegant older-style
motorboat with electric drive, manufactured by Kyosho
as an ARR (almost-ready-to-run) kit. Epoxy-treated
wood bonded to an ABS plastic hull has created a
wonderfully luxurious wood finish.*

LEFT *One of the most
spectacular radio-
controlled models is this
1:40 scale replica of the
famous World War II
U47 German
submarine. Produced by
Robbe, the model has
many authentic details
taken from the plans of
the real submarine. It
has a pressure hull for
underwater running
and is 68in (1.7m)
long. The hull breaks
into two parts for
carrying. The hull and
main components are in
ABS plastic, and the
model has full diving
and trimming
performance with up to
six-channel radio
control.*

RESOURCES

If you are a newcomer to radio-controlled models, you may want to try out a variety of different models, such as cars, gliders and powerboats. Most modelers, however, end up specializing in one area. Whichever type of model you choose, you can be sure that there will be exciting new kits and accessories coming out each year to stimulate your interest. Many resources are available to help you keep up with the latest technological developments and the most recent model releases.

MAGAZINES

Reading specialist radio-control magazines is one of the best ways to keep up with the latest developments and innovations. In the United States and Great Britain, in particular, there are many magazines devoted to specific model types, such as boats, aircraft, and cars. In addition, some general modeling magazines include radio-control sections. Most magazines appear monthly, but some may be bi-monthly or quarterly. The club and association news sections in these magazines are a good place to start, if you are interested in meeting other modelers and getting involved in racing and other model events. Check at your local hobby dealer or newsstand for current availability, since titles change from time to time.

Boat and Ship Modeler
Model Builder (radio-controlled aircraft)
Radio Control Model Cars
PO Box 2459
Capistrano Beach
California 92624–0459
USA
(714) 496 5411

Radio Control Buyer's Guide
(published annually)
Kalmbach Publishers
21027 Crossroads Circle,
Waukesha
Wisconsin 53187
USA
(800) 446 5489

Aviation Modeller International
Model Activity Press
5 George Street
St Albans
Hertfordshire AL3 4ER
UK
(1727) 840010

Radio Modeller
RCM&E (Radio Control Models and Electronics)
Model Boats
Radio Control Boat Modeller
R/C Model Cars
Aero Modeller
Nexus
Boundary Way
Hemel Hempstead
Hertfordshire HP2 7ST
UK
(1442) 66551

RC-Freizeit (RC-F)
(Also distributed in Austria, Switzerland, Holland, and Italy)
Postfach 102224
86012 Augsburg
Germany
(821) 39268

Allt om Hobby
Box 90 133
120 21 Stockholm
Sweden
(89) 99333

EXHIBITIONS AND EVENTS

A number of local and national exhibitions take place throughout the year in most countries. Manufacturers and clubs have stands at these events, and kits and accessories can usually be purchased. It is also worth attending to see the demonstrations, presentations, and competition events. Racing radio-controlled models is a popular event—check with your local hobby dealer or specialist magazine for details of upcoming model-boat regattas, aircraft events, and car competitions.

BOOKS

There are a number of specialist books on specific aspects of radio-controlled models. However, because of their specialist nature they may be difficult to find in general bookstores—check with your local hobby dealer or public library. The *Radio Control Guide Book,* produced by Tamiya, is packed with useful information, particularly regarding car and buggy racing. Aimed specifically at owners of Tamiya kits, the book is updated regularly.

VIDEOS

Instructional videos demonstrating techniques for making and operating radio-controlled models are a recent innovation. The videos produced by the leading American model manufacturer Great Planes are particularly useful for understanding the complex techniques needed for building model airplanes, such as kit assembly, and covering and finishing. In addition, some videos are included free with more complicated kits.

FLIGHT SIMULATORS

A good way to hone your flying skills is to try computer flight simulators. These allow you to fly a model aircraft in simulation and are useful for finding out more about a model's movements and capabilities. Flight software, mostly for Windows 95, is produced by several model manufacturers such as Great Planes and Kyosho.

MANUFACTURERS AND DISTRIBUTORS

All the major manufacturers and distributors, such as Tamiya, Kyosho, Ripmax, and Robbe, produce detailed catalogs each year; they may also send out newsletters or brochures. These superb catalogs list all currently available models and accessories. Small specialist companies also produce catalogs. Ask at hobby stores for these catalogs, or call the manufacturer.

Many leading manufacturers also have Internet websites which can be a good source of information. In addition, some manufacturers' catalogs are available on CD-Rom.

Great Planes Model Distributors
(most international models)
PO Box 9021
Champaign
Illinois 61826–9021
USA
(217) 398 6300

Nikko America Inc
2801 Summit Avenue
Plano
Texas
USA
(972) 422 0838

Precision Model Distribution
(Wedico)
155 West 7th Place
Mesa
Arizona 85201
USA
(602) 655 9888

Tamiya America
2 Orion
Aliso Viejo
California 92656–4200
USA
1-800-Tamiya-A (toll-free)

Nikko UK
7 Little Mundells
Welwyn Garden City
Hertfordshire AL7 1EW
UK
(1707) 377771

Riko Ltd (Tamiya)
13–15a High Street
Hemel Hempstead
Hertfordshire HP1 3AD
UK
(1442) 261 721

Robbe UK
51 Sapcote Road
Burbage
Hinckley
Leicestershire LE10 2AS
(1455) 635151

Ripmax plc (Kyosho, Graupner, Great Planes)
Ripmax Corner, Green Street
Enfield
Middlesex EN3 7SJ
UK
(181) 804 8272

Graupner GmbH
Henriettenstrasse 94–96
73230 Kirchheim/Teck
Germany
(911) 8168629

Robbe Modellsport GmbH
Metzloserstrasse 36
36355 Grebenheim
Germany
(6644) 87137

Wedico Truck-Modelle (model trucks and accessories only)
Dahler Strasse 72
42389 Wuppertal
Germany
(202) 26 6000

Kyosho Corporation
153 Funako
Atsugi
Kanagawa 243
Japan
(462) 291511

Nikko Co Ltd
5-15-15 Kameari
Katsushika-ku
Tokyo 125
Japan
(3) 3620 3555

Tamiya Plastic Model Co
628 Oshika
Shizuoka 422
Japan
(54) 285 5187

CLUBS AND SOCIETIES

If you are interested in learning more about radio-controlled models or want to try model racing, it is a good idea to join a club or association. You can usually find contact addresses in model magazines. Joining a club is particularly useful if you are interested in aircraft or helicopters, because it is easier to learn operating skills for these models from an experienced training "buddy" than on your own. Some clubs may also run training schemes aimed specifically at newcomers to radio control.

Academy of Model Aeronautics
(model airplanes)
5151 East Memorial Drive
Muncie
Indiana 47302
USA
(317) 287 1256

American Model Yachting Association
2793 Shellwock Court
Columbus
Ohio 43235
USA
(614) 457 1185

American Power Boat Association
29624 Lahana Way
Fremont
California 94538
USA
(510) 656 7072

International Electric Drag Racing Association
115 Kerr Street
Clinton
North Carolina 28328
USA
(910) 592 9489

International Model Power Boat Association
5855 S. State Road
9 Fountaintown
Indiana 46130
USA
(317) 861 3701

Model Powerboat Association
2 Eglington Road
London E4 7AN
UK
0181 524 7840

British Model Flying Association
31 St Andrews Road
Leicester LE2 8RD
UK

British Radio Car Association
18 Cedar Avenue
Enfield
Middlesex
UK

Model Yachting Association (UK)
115 Mayfield Avenue
London N12 9HY
UK

International Yacht Racing Union (UK)
60 Knightsbridge
London SW1X 7JK
UK
(0171) 235 6221

GLOSSARY

ABS
Flexible type of plastic used for many model bodies

Aerial
Wire or metal rod used to facilitate transmission or reception of radio signals

Airfoil section (US), aerofoil section (UK)
Shaped cross-section of a wing that generates lift in a flow of air

Aileron
Moving control surface on wing

Airbrake
Device (usually on a glider) to be raised to reduce speed

Air-operated retract
Air-charged device for large-scale flying models that retracts the undercarriage. Sprung retracts are also available.

Airscrew
Airplane propeller

Allen key
Tightening tool, or wrench, much used in model making

ARTF
Almost-ready-to-fly models

ARTG
Almost-ready-to go models

ARTR
Almost-ready-to-run models

Autopilot
Unit that acts as a failsafe to return controls to "neutral" if there is a control malfunction

Ball bearing
Superior type of bearing, mostly found in electric motors and car mechanisms

Balsa
Very light, but strong wood much used for radio-controlled model aircraft; it comes in many forms including strip and sheets

Battery charger
Essential unit for recharging ni-cad batteries from the mains or car battery. Some are dedicated to one type of battery, others have an adjustable output current for different types of battery

Biplane
Type of aircraft with upper and lower wings

Brush
Key part of an electric motor

Buddy system
Flying training method using two matched transmitters

Bungee
Rubberized cord used to launch model (and real) gliders

Bungee launch
Taking off using a bungee cord

Cardan shaft
Flexible drive shaft used in transmission systems; may be sprung, jointed, or telescopic, to accommodate movement of wheels, prop, etc

Clutch
Device used on engines and transmissions to engage the drive; often automatic to engage when a certain speed is reached

Coaming
Raised edge on boats to keep out water

Commutator
Key component on the shaft of an electric motor

Condensor
Essential component of an electric motor

Coupling
Used in conjunction with drive shafts; often flexible and commonly found in power boat transmissions

Covering(s)
Tissue or self-adhesive plastic film used to cover airplane surfaces

Crystal
Essential component that can be changed and matched in transmitter and receiver, to ensure each model present operates on a different channel; it is often abbreviated to Xtal; note that a crystal in a transmitter is marked TX, and one in a receiver RX

Cyanoacrylate
Quick bonding glue, or "superglue," usually abbreviated to CA or CAN

Dampers
Oil-filled telescopic arms, or rubber pads, used to "soften" supensions on some transmission systems

Decals
Decorative markings used on all models. Pressure sensitive, waterslide, varnish, and self-adhesive decals are available; the latter are the most common

Differential
Gear enabling shafts to revolve at different speeds, mostly in model cars and trucks

Disk (or disc) brake
Type of brake found in more advanced radio-controlled cars

Ducted fan
Electric motor mount with propeller inside nacelle or tube

Electronic speed controller
Unit allowing proportional speed control of electric motor with minimum power loss

Elevator
Moving control surface on stabilizer (tailplane), affecting climb or dive movement of airplane

Epoxy
Material in liquid or putty form that sets to give a light, but durable surface

Fiberglass
Polyester-based hard material, also known as GRP (glass reinforced plastic), that can be molded to shape; it is used in some boat and yacht hulls, and in some larger model airplanes

Fin
Upright fixed portion of airplane tail

Firewall
Fireproof bulkhead fitted to some larger IC-powered models

Flybar paddle
Device to counterbalance operation of main motor blades of helicopter

Flywheel
Weighted wheel in some engine and transmission systems, for smoother operation

Folding prop
Collapsible propeller fitted to model powered gliders

Frequency/frequency band
Frequency bands are allocated to radio-controlled models; each model operates on a dedicated channel within the frequency band range

Frequency pennant
Numbered or colored flag attached to transmitter aerial to indicate operating frequency

Fuselage
Main body of airplane

Fuel-proofer
Fireproof coating applied to IC-engined models

Governor
Precision device for controlling engine speed; found in some advanced systems

Gunwale
Upper edges of ship's sides

Gyro
Sensor unit with spinning weight that corrects rotational movement in helicopters; also used in some large aircraft models to give added stability

Hard chine
Shallow V-section lower hull formation of high-speed powerboats

Heatsink
Heat-conductive mounting or pad, generally made of metal, to which an IC engine is fitted

High-wing aircraft
Aircraft with wings set above fuselage

Hinge(s)
Very small light versions of conventional hinges, mostly made of plastic, for attaching moving control surfaces

Horns
Devices used to attach the operating linkages from the servos to the control surfaces or other moving parts; they are mostly made of plastic or metal and are sometimes adjustable

Hot-up accessories
Accessories for boosting the power and performance of your model

Hydrojet
Propulsion system on modern fast boats enabling movement by sucking in water and ejecting it to rear via an impeller

Hydroplane
Power boat driven by aircraft motor and propeller on hull top

IC engine
Internal combustion engine, also known as gas engine, glow engine, or nitro engine

Ladder-frame chassis
Traditional chassis structure resembling a ladder

Lexan
Type of polycarbonate, strong, flexible plastic used in many model superstructures, especially cars

Linkages
Metal or plastic rods used to transmit the movements from servos to control surfaces or other moving parts (eg. throttles, steering arms); also available in flexible form (known as "snakes") for use where the run between servo and moving part is not direct, as, for example, in the wing of an aircraft, where ailerons need to be controlled; also used are extenders for joining two linkages for a long run and links for right-angle connections or changes of direction; note that linkages are also known as pushrods, or control rods

Low-wing aircraft
Aircraft with wings attached at bottom of fuselage

Manifold
Exhaust pipe of an IC engine

Micro motor
Very small, light electric motor for very small models; also used to provide drives when space is limited

Monocoque
New-type one-piece chassis

Monowheel
Single landing-wheel fitted under fuselage of larger model gliders (also used on real gliders)

Muffler (US), Silencer (UK)
Exhaust silencer, usually including a baffle chamber

Nacelle
Streamlined housing for motor or other external airplane fitting

Nicad (nickel-cadmium)
High-power battery used for radio-controlled models, usually in pack form in a variety of power ratings; can be recharged with a battery charger; sometimes written as Ni-Cad, NC, NiCd, or nicd

Nose gear
The front wheel of an airplane's tricycle undercarriage

Obechi
Light, but strong wood used for some model airplanes

Phillips screwdriver
Specialized crosshead blade tool for Phillips crosshead screws, often used in assembly of radio-controlled models

Power glider
Glider fitted with a small electric motor to assist in gaining height

Power unit
General term for electric motor or IC engine used to propel a model

Proportional Control
Modern method of controlling a model, whereby the amount of movement imparted to the control stick is directly related to the movement of the controls. For example, moving the stick exactly halfway left will cause the rudder on a model boat to move exactly halfway left.

Prototype
The full-sized original

Pushrod
See linkage

Receiver
Unit that picks up the radio signal from the transmitter, and directs the information to the servos

Rotor head
In helicopters, assembly holding the rotor and the controls that changes its angle, etc

RTF
Ready-to-fly model

RTR
Ready-to-run model

Sail control unit (US), Sail winch (UK)
Servo hauling yacht sheets in and out

Semi-scale airplane
Radio-controlled model closely resembling a genuine airplane, but with simplification or minor changes to make flying easier

Servo
Motorized device with a rotor that is linked by a pushrod to the function being controlled (e.g. steering); note that one servo is needed for each function

Shoulder wing
Wings on a monoplane attached at top of fuselage

Slicks
Racing tires on cars

Slot car
Small-scale racing car guided by a slot in the track; sometimes they may be adapted for radio control

Spinner
Center pivot, usually with a streamlined fairing, of an airplane propeller

Spoiler
Fairing, or blade, for deflecting airflow or keeping balance

Sponson
Extension from a main structure to provide extra support or to stabilize a boat

Sports plane
Radio-controlled model airplane used for racing or aerobatics

Stabiliser (US), Tailplane (UK)
Horizontal, fixed section of tail assembly

Throttle
Control unit for motor/engine speed

Trailing edge
Rear edge, usually of an airplane wing or tail

Trainer
Widely available radio-controlled model airplane for beginners; usually high-wing, single-engine type

Transmitter
Key unit that generates and sends the signal controlling model movements; signal is AM (amplitude modulation) or FM (frequency modulation), as with domestic radio, and transmits on designated frequencies

Transom
Squared off stern face of power boat or yacht

Vac-form plastic
Molded plastic

Yaw
Swinging movement left or right, sometimes controlled, but often involuntary, on forward moving model

INDEX

ACKNOWLEDGEMENTS

The author would like to thank the modelers who made models specially for this book. The models featured in this book were made by:

Chris Ellis
Tom Ellis
Dave Kemp
Malcolm Lowe
Justin Rabbetts
Michael Smith

The author would also like to thank the model companies and distributors, who generously supplied kits, accessories, materials, photographs, and information for inclusion in this comprehensive reference book.

Special thanks to:
Graupner Modellbau (Anny Kreyscher)
Great Planes Model Distributors' Company (Mark Forcier)
Kyosho Corporation
Nikko Co Ltd/Nikko UK Ltd (Graham R. Stevens)
Richard Kohnstam Ltd/Riko (Peter Binger, Glyn Pearson)
Ripmax plc (Ian Richards, Julia Miles)
Robbe Modellsport (Roya Toloui, Mike Atkinson)
Tamiya Plastic Model Co (T. Hirayama)
Wedico GmbH (Marion von Zastrow)

The author is also grateful to Malcolm Lowe for his help with the aircraft and powerboat sections, and Tom Ellis for his assistance with the car and truck sections.

Additional photographs by:
Chris Ellis
Malcolm Lowe
Andy Barnes